7th Grade Mathematics

This Test Review Booklet was designed for Grade 7 Mathematics Assessment Test. It provides examples of the format and types of questions that may be on the actual test as administered by the State Education Department. We have separated our review tests into two sections:

    Part 1 - Multiple Choice Section
    Part 2 - Short / Extended Response Section

The actual test has three books, administered over three days.

Book 1: 26 multiple choice questions

Book 2: 25 multiple choice questions

Book 3: 10 short or extended response questions

Each section should take approximately 60 minutes.

Each student should have the following items made available to them during the test:

    -Ruler
    -Protractor
    -Calculator
    -Mathematics Reference Sheet

Students must be able to use a scientific calculator. These calculators must include normal arithmetic operations, decimal, change sign, parentheses, square root and Pi functions. Graphing calculators or those with problem solving, programming, place value or inequality solution facilities are not allowed.

For a complete description of restrictions involving calculator usage, see the NY State Education website: www.nysed.gov

© 2016, Topical Review Book Company, Inc. All rights reserved.
P. O. Box 328
Onsted, MI. 49265-0328

This document may not, in whole or in part, be copied, photocopied, reproduced, translated, or reduced to any electronic medium or machine-readable form without prior consent in writing from Topical Review Book Corporation

Question Distribution

The questions in the practice exams have the same approximate distribution as that described by the NYSED Elementary, Middle, Secondary, and Continuing Education guidelines. They are:

Ratios and Proportional Relationships	20-30%
The Number System	15-25%
Expressions and Equations	30-40%
Geometry	5-15%
Probability and Statistics	10-20%

Special Thanks To:

Luke Masouras - Examgen Inc.

Syracuse, NY • www.EXAMgen.com

For providing technical guidance and development of the test questions

Published by
TOPICAL REVIEW BOOK COMPANY
P. O. Box 328
Onsted, MI. 49265-0328

Grade 7 Mathematics Reference Sheet

CONVERSIONS

1 inch = 2.54 centimeters	1 kilometer = 0.62 mile	1 cup = 8 fluid ounces
1 meter = 39.37 inches	1 pound = 16 ounces	1 pint = 2 cups
1 mile = 5,280 feet	1 pound = 0.454 kilogram	1 quart = 2 pints
1 mile = 1,760 yards	1 kilogram = 2.2 pounds	1 gallon = 4 quarts
1 mile = 1.609 kilometers	1 ton = 2,000 pounds	1 gallon = 3.785 liters
		1 liter = 0.264 gallon
		1 liter = 1,000 cubic centimeters

FORMULAS

Triangle	$A = \frac{1}{2}bh$
Parallelogram	$A = bh$
Circle	$A = \pi r^2$
Circle	$C = \pi d$ or $C = 2\pi r$
General Prisms	$V = Bh$

7th Grade Mathematics

TABLE OF CONTENTS

EXAM		PAGE
Test 1	Part 1:	1
	Part 2:	12
Test 2	Part 1:	17
	Part 2:	29
Test 3	Part 1:	33
	Part 2:	44
Test 4	Part 1:	48
	Part 2:	59
Test 5	Part 1:	63
	Part 2:	74
Test 6	Part 1:	78
	Part 2:	89

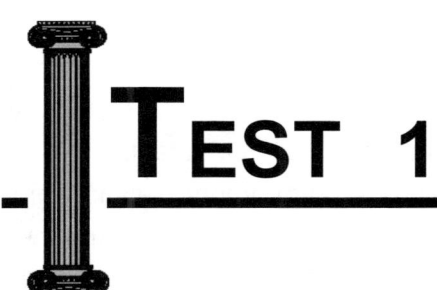

TEST 1

PART 1

1 Write "$33,000 earned in 1 year" as a unit rate in months.

   A $2,570/month
   B $1,300/month
   C $5,500/month
   D $2,750/month   1 _____

2 What is "78 radishes planted in a 13-foot long row" written as a unit rate per inch?

   A 72 radishes/inch
   B 0.5 radish/inch
   C 6 radishes/inch
   D 2 radishes/inch   2 _____

3 What is "1,017 square feet of lawn mowed in 9 minutes" written as a unit rate per hour?

   A 6,780 square feet mowed/hour
   B 113 square feet mowed/hour
   C 61,020 square feet mowed/hour
   D 753 square feet mowed/hour   3 _____

4 Write "$157.50 earned in 15 hours" as a unit rate per 40-hour workweek.

   A $945/week
   B $10.50/week
   C $420/week
   D $600/week   4 _____

5 The graph shows below the distance Marissa travels on her bike during a forty-minute period.

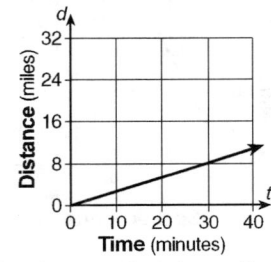

What is the approximate unit rate of Marissa's biking speed?

   A $3\frac{3}{4}$ mi/min
   B 240 mi/min
   C $\frac{1}{4}$ mi/min
   D $\frac{1}{2}$ mi/min   5 _____

6 The number of pieces of mail processed by a machine in the post office is directly proportional to the number of minutes that the machine runs. The machine processes 2,700 pieces of mail in 60 minutes of continuous running. What is the speed of this machine expressed as a unit rate?

A 45 pieces/minute
B 45 pieces
C 45 pieces/hour
D 45 pieces/second   6 _____

7 Which one of the following ratios has the same unit rate as the line graphed below?

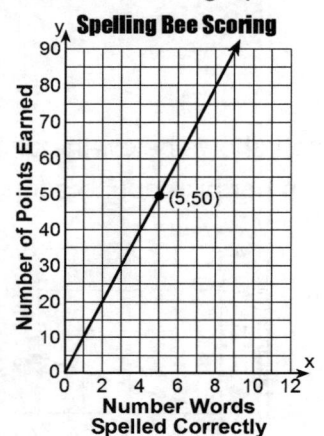

A $\frac{55}{15}$
B $\frac{40}{4}$
C $\frac{8}{85}$
D $\frac{7}{70}$ 7 _____

8 A state income tax is 9% of any amount earned over $15,000. What state income tax does a person earning $24,500 pay?

A $1,350
B $855
C $2,205
D $650   8 _____

9 Amity and Beth are traveling in a taxi. The graph below shows a proportional relationship between the time they rode in the taxi and the cost of the taxi ride. Which expression can be used to find the unit rate in dollars?

A 3 + 4.5
B $\frac{9}{6}$
C $\frac{6}{9}$
D $\frac{0}{0}$ 9 _____

10 A person gets a $2,000 home improvement loan from a bank for six months at an annual interest rate of 12%. What is the simple interest due on the loan?

A $140
B $210
C $120
D $240   10 _____

11 Approximately how long will it take $200 to double to $400 when the annual interest rate is $8\frac{1}{2}$%?

A 6.25 years
B 11.75 years
C 1.75 years
D 8.5 years   11 _____

12 Mark bought an MP3 player for $275. If the sales tax is 8.25%, what was the total cost of the MP3 player?

A $283.25
B $266.75
C $297.69
D $252.31   12 _____

13 The bill for a meal at a restaurant is $105.95. If the customer leaves a 20% tip, what is the total cost of the meal, with tip?

A $127.14
B $21.19
C $115.93
D $145.00   13 _____

14 If the check at a restaurant totals $115.99 and the customer leaves a 15% tip, what is the total cost of the meal?

A $133.39
B $123.39
C $17.40
D $143.39   14 _____

15 Add: +18 + (−25)

A +7
B +43
C −43
D −7   15 _____

16 Add: −42.67 + (+19.4) + (−1.2)

A 24.87
B 24.47
C −22.07
D −24.47   16 _____

17 Add: $+3\frac{1}{4} + (-2\frac{1}{2})$

A $-1\frac{1}{4}$
B $5\frac{3}{4}$
C $\frac{3}{4}$
D $-\frac{3}{4}$ 17 _____

18 What number makes the subtraction sentence $\square$ − −8.73 = 22.09 true?

A 13.36
B 30.82
C −30.82
D −13.36   18 _____

Test 1 – Part 1

19 $(-3\tfrac{1}{2}) - (-8\tfrac{2}{3}) =$

- A $\quad -12\tfrac{1}{6}$
- B $\quad -5\tfrac{1}{6}$
- C $\quad 5\tfrac{1}{6}$
- D $\quad 12\tfrac{1}{6}$

19 _____

20 Which decimal is equal to $\tfrac{1}{8}$?

- A $\quad 3.6$
- B $\quad 0.125$
- C $\quad 8.0$
- D $\quad 0.36$

20 _____

21 Which decimal is equal to $\tfrac{7}{20}$?

- A $\quad 0.35$
- B $\quad 9.0$
- C $\quad 2.86$
- D $\quad 0.9$

21 _____

22 Which decimal is equal to $\tfrac{6}{25}$?

- A $\quad 8.5$
- B $\quad 0.24$
- C $\quad 0.85$
- D $\quad 4.167$

22 _____

23 Claude lives $1\tfrac{4}{5}$ miles from school. If he jogs $\tfrac{1}{2}$ this distance every day, how far does he jog daily?

- A $\quad \tfrac{2}{5}$ mile
- B $\quad 1\tfrac{2}{5}$ mile
- C $\quad \tfrac{4}{5}$ mile
- D $\quad \tfrac{9}{10}$ mile

23 _____

24 Three girls want to share $5\tfrac{1}{4}$ pounds of taffy equally. How much taffy should each girl get?

- A $\quad 1\tfrac{1}{3}$ pounds
- B $\quad 2\tfrac{1}{3}$ pounds
- C $\quad 1\tfrac{5}{8}$ pounds
- D $\quad 1\tfrac{3}{4}$ pounds

24 _____

25 A car used $12\tfrac{3}{10}$ gallons of gasoline on a 369-mile trip. How many miles did this car travel on one gallon of gasoline?

- A $\quad 24$
- B $\quad 26$
- C $\quad 28$
- D $\quad 30$

25 _____

26 How many bags of flour weighing $2\frac{1}{2}$ pounds each can be made from a sack of flouring weighing 45 pounds?

    A   19
    B   18
    C   17
    D   16   26 _____

27 How many lengths of $1\frac{7}{8}$ inches can be cut from a wooden dowel measuring $7\frac{1}{2}$ inches?

    A   5
    B   6
    C   4
    D   3   27_____

28 During the month of August, Nick worked $\frac{2}{3}$ as many hours as he did during the month of July. If he worked 24 hours in August, how many hours did he work in July?

    A   32
    B   16
    C   38
    D   36   28 _____

Questions 29 through 34 refer to the following:

Simplify the given expression:

29 $-13m + m =$

    A $-14m$
    B $-12m$
    C $14m$
    D -12 29 _____

30 $14x + 9 + 6x + 4 =$

    A $20x + 13$
    B $33x$
    C $20x^2 + 13$
    D $33x^2$ 30 _____

31 $5x - 3x + 5 =$

    A $2x + 5$
    B $-2x + 5$
    C $2x^2 + 5$
    D $7x$ 31 _____

32 $7y^2 - 3y^2 =$

    A   4
    B $-4y^2$
    C $4y^2$
    D $4y^4$ 32 _____

33 $-9x^2 - 3x^2 - 4x^2 =$

A $-2x^2$
B $16x^2$
C $-16x^2$
D $-16x^6$

33 _____

34 $8x^3 - 5x^3 =$

A $3x^3$
B $8x^3$
C 3
D $-3x^3$

34 _____

35 Jeremy is a surveyor. He earned x dollars per week for two weeks of job training. He earned 5% more on his first full week of actual work. What expression represents the total amount of money Jeremy earned during these three weeks?

A $1.05x$
B $1.05x + 2$
C $3.50x$
D $3.05x$

35 _____

36 Annie's Room, a children's furniture store, is having a sale. They advertised that every item in the store is 35% off the regular price. What expression represents the cost, in dollars, of an item that was marked x dollars?

A $0.35x$
B $x - 0.35$
C $1 - 0.35x$
D $x - 0.35x$

36 _____

37 How is the expression below rewritten using the Distributive Property?

$$\frac{1}{2}k + \frac{1}{4}k$$

A $\frac{1}{2}k(1 + \frac{1}{8})$
B $\frac{1}{4}k(\frac{1}{4} + 1)$
C $\frac{1}{4}k(2 + 1)$
D $\frac{1}{4}k(\frac{1}{2} + 1)$

37 _____

38 The value of a house this year is $114,000. This is three times its value eighteen years ago. Find its value eighteen years ago.

A $342,000
B $38,00
C $28,00
D $48,00

38 _____

Page 6 | Test 1 – Part 1 | Copyright © 2016 Topical Review Book Company

39 Solve the equation for the given variable:

$$3y + 29 - 5y = 43$$

- A −12
- B −7
- C −36
- D −4

39 _____

40 A banquet hall charges $100 for the use of its dining room and $7.50 a plate for each dinner. The Civic Club hosts a dinner party at the banquet hall and charges $10.00 a plate, but also invites ten nonpaying guests. If each person orders one plate, how many paying persons must attend for the Civic Club to collect the amount needed to pay the total cost of the dinner exactly?

- A 70
- B 50
- C 30
- D 40

40 _____

41 Four times as many girls as boys participate in chorus. If there are a total of 140 girls and boys, how many girls are in the chorus?

- A 70
- B 112
- C 105
- D 28

41 _____

42 A woman invested x dollars at an interest rate of 9%. The annual income from the investment is $400. What equation expresses this relationship?

- A $0.09x = 400$
- B $0.09(400) = x$
- C $\frac{x}{400} = 0.09$
- D $0.9x = 400$

42 _____

43 Larry has 7 more dimes than nickels, for a total value of $1.45. If n represents the number of nickels, what equation could be used to find the number of nickels Larry has?

- A $15(n + n + 7) = 145$
- B $5n + 10(n + 7) = 145$
- C $5n + 5(n + 7) = 145$
- D $n + (n + 7) = 145$

43 _____

44 If the actual distance between two locations is 60 miles and the map scale is 2 inches = 30 miles, what is the distance between the two locations on the map?

- A 4 inches
- B $\frac{1}{2}$ inch
- C 1 inch
- D 2 inches

44 _____

Test 1 – Part 1

45 The distance between two locations on a map is $4\frac{1}{8}$ inches. If the map scale is 2 inches = 40 miles, what is the actual distance between the two locations?

A 165 miles
B 41.25 miles
C 100 miles
D 82.5 miles   45 _____

46 Lance is drawing a triangle. He draws one side that is 8 inches long and another side that is 5 inches long. Which one of the following could be the third side?

A 1 inch
B 3 inches
C 2 inches
D 4 inches   46 _____

47 Which of the following two-dimensional faces can be formed from a cross-section of a cube?

A rectangle, only
B square and rectangle, only
C square, rectangle, and hexagon
D square, only   47 _____

48 The diagram below represents a cube sliced through all six faces.

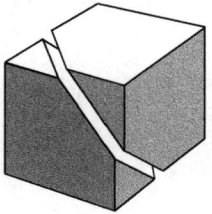

What is the sum of the interior angles of the two-dimensional surface resulting from this cross-section?

A 360°
B 180°
C 720°
D 540°   48 _____

Question 49 refers to the following:

Given the formulas for the circumference and area of a circle below.

Circumference = $2\pi r$
Area = πr^2

49 What is the circumference of a circle whose area is 25π?

A 10π
B 50π
C 5π
D 25π 49 _____

Questions 50 and 51 refer to the following:

Given the formulas for the circumference and area of a circle below.

Circumference = 2π r
Area = π r²

50. What is the circumference of the circle pictured below?

- A 18π
- B 8π
- C 81π
- D 36π

50 _____

51. What is the circumference of the circle pictured below? [Use π = 3.14.]

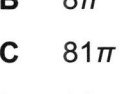

- A 37.68
- B 34.54
- C 31.40
- D 36.11

51 _____

52. If the measures of two complementary angles are in the ratio 1:5, the measure of the larger angle is

- A 144°
- B 150°
- C 72°
- D 75°

52 _____

53. In the accompanying diagram, line $\overleftrightarrow{RS}$ and line $\overleftrightarrow{TU}$ intersect at point W.

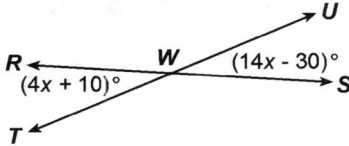

If m∠RWT = (4x + 10)° and m∠UWS = (14x − 30)°, what is the value of x?

- A 2
- B 8
- C 4
- D 2.5

53 _____

54. The Meyers family decided to cover their outdoor deck with Astroturf. The deck is in the shape of a rectangle with dimensions of 24 feet by 16 feet. The Astroturf to be used costs $3.20 per square foot. What will it cost to cover the deck?

- A $614.40
- B $256.00
- C $1,344.00
- D $1,228.80

54 _____

55. If a certain mixture costs $1.85 per cubic foot, what would it cost to fill a rectangular container measuring 36 ft by 12 ft by 2 ft with that mixture?

- A $92.50
- B $1,598.40
- C $799.20
- D $864.00

55 _____

PART 2

56 How many gallons of water (to the nearest tenth) will fill an aquarium that is 10.5 in. wide, 27 in. long, and 13.5 in. high? [1 gallon = 231 in^3.] [Round to the nearest tenth.]

Show your work:

Answer:_____

57 Carly is wrapping a present for her friend's birthday. She could only find wrapping paper that is 25 inches wide by 38 inches long. Does she has enough paper to wrap a box measuring 9 inches by 16 inches by 10 inches.

Show your work and explain your answer on the lines below:

58 Zoe is recovering her couch ottoman with the dimensions in both inches and feet shown below.

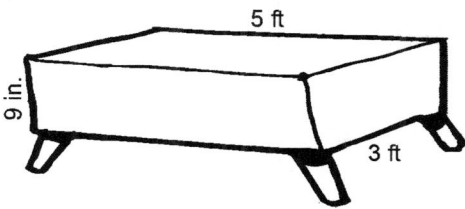

How many square feet of fabric does she need to purchase to cover the 5 sides, including an extra 2 inches on each bottom edge to fold underneath and staple to the wood base?

Show your work:

*Answer:*_____ft²

59 A carbonated beverage company needs to report the mean ages of their employees in two different departments and compare them. The data are listed in the table.

Compare the mean ages of the two departments and describe the difference between them.

Show your work. Round to the nearest whole number:

Ages of Employees

Sales	Delivery
23	18
34	21
35	52
67	24
41	33
20	47
30	31

Sales Mean: _____ years

Delivery Mean: _____ years

Difference: _____ years

60 Find the average of the numbers in the stem and leaf plot to the nearest whole number.

Show your work:

Stem	Leaf
4	9 9
5	2 3 3 3 5 7 8 8 8
6	0 0 4 4

KEY: 6|4 means 64

Answer:_____

61 A spinner is divided into 3 equal sections labeled A, B, and C.

Predict how many times out of 180 spins that the spinner will most likely stop on a vowel.

Show your work:

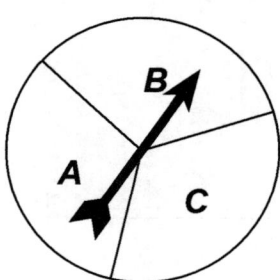

Answer:_____

62 The spinner is divided into 8 equal sections.

Predict how many times out of 320 spins that the spinner shown will most likely stop on a vowel.

Show your work:

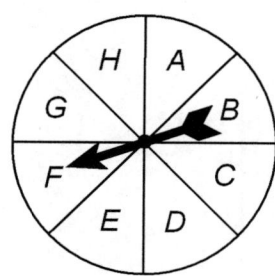

Answer:_____

63 Julie tosses a coin 50 times to see how many times tails lands face up and how many times heads lands face up. She recorded her results in the table below.

Heads	Tails
27	23

Part A
Before the experiment, how many times would you have expected to see tails land face up?

*Answer:*_____

Part B
Based on her results, what is the experimental probability of seeing tails land face up?

*Answer:*_____

Part C
In theory, how many times should you expect to see tails land face up when the coin is tossed 200 times?

*Answer:*_____

Part D
Based on her results, how many times would you expect to see tails land face up when the coin is tossed 200 times?

*Answer:*_____

64 **Part A**

Write the sample space for the outcomes of the toss of two six-sided die.

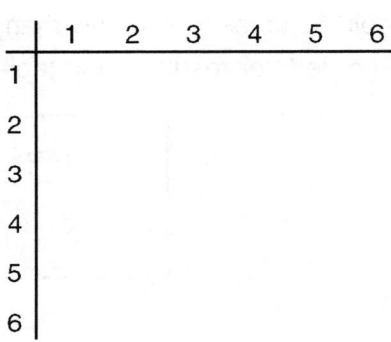

Part B

If order is important, how many outcomes are possible?

*Answer:*_____

Part C

If order is not important, how many outcomes are possible?

*Answer:*_____

65 Stefan and Gabe are playing a game that uses a standard 6-sided numbered die and a 10-sided lettered die labeled *A* through *J*. In the space below, create a table, organized list, or tree diagram to determine all possible outcomes.

Using the model you have drawn, what is the probability of rolling one of the first six letters of the alphabet and a number *greater than* 4?

*Answer:*_____

Test 2

PART 1

1 Which of the following is a valid proportion?

A $\dfrac{5}{9} = \dfrac{30}{45}$

B $\dfrac{1}{2} = \dfrac{30}{50}$

C $\dfrac{2}{3} = \dfrac{44}{66}$

D $\dfrac{5}{8} = \dfrac{35}{64}$

1 _____

2 Which of the following graphs illustrates a proportional relationship?

A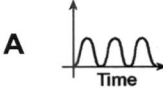

B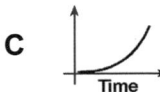

C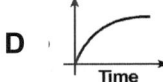

D

2 _____

3 What percent of the figure below is shaded?

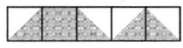

A 60%

B 50%

C 75%

D 40%

3 _____

4 For which survey question below would it be useful to include a graph to display the resulting answers?

A Can students have pizza delivered at school?

B Do students like pineapple on their pizza?

C What are the students' favorite pizza toppings?

D How many students have ever made pizza?

4 _____

5 Which one of the following statements represents a relationship equal to 0?

A investing $100 and earning $100

B earning $7 and spending $7

C getting $10 change from $20

D earning $10 and saving $10

5 _____

Test 2 – Part 1

6 During a game, Myra was dealt the hand below, containing cards labeled 1 through 6.

If another player randomly chooses a card from Myra's hand, what is the probability that the card is a 3?

A $\frac{1}{3}$

B $\frac{3}{15}$

C $\frac{2}{3}$

D $\frac{2}{15}$ 6 _____

7 A customer buys a refrigerator for $650 and must pay a sales tax of 6% of the cost. What is the total cost of the refrigerator?

A $589
B $108.33
C $679
D $689   7 _____

8 Find the product of the given set of numbers:
(+15)(+6)

A 90
B −85
C −90
D 85   8 _____

9 Simplify the given expression:
$a + a =$

A a^2
B $2a$
C a
D $2a^2$ 9 _____

10 The first snow of the season is falling at a rate of 4 inches an hour. Which of the following graphs best shows a proportional relationship between the depth of the snow on the ground and the amount of time that has passed?

A 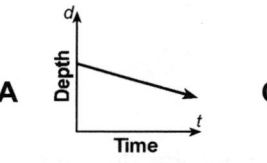   C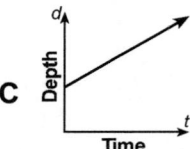

B 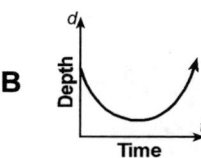   D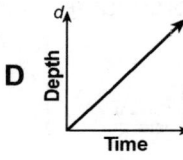

10 _____

11 A chef cooked a roast weighing $12\frac{1}{3}$ pounds. After $3\frac{2}{3}$ pounds of fat were trimmed, the roast was cut into $\frac{1}{3}$ pound servings. How many servings were cut from the roast?

A 24
B 29
C 18
D 26   11 _____

12 What is $2\frac{1}{3}$ divided by $-1\frac{1}{3}$?

A $-\frac{7}{4}$

B $\frac{7}{4}$

C $-\frac{7}{9}$

D $\frac{7}{9}$

12 _____

13 Which one of the following expressions has the least value?

A $(3 - 7) \times (-3) - 6$

B $3 - 7 \times (-3) - 6$

C $(3 - 7) \times (-3 - 6)$

B $3 - (-3 - 6 \times 7)$

13 _____

14 The two acute angles in an isosceles right triangle must measure

A 30° and 60°

B 45° and 45°

C 35° and 55°

D 40° and 50°

14 _____

15 The diagram below shows a right circular cone with four possible cutting paths, A, B, C, and D.

Cutting along which lettered path in the given diagram will result in a dilation of the unshaded surface?

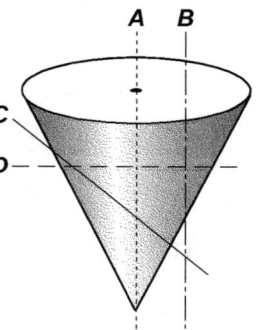

A A

B B

C D

D C

15 _____

16 Given $\angle CBD = 20°$, what is the measure of $\angle ABD$.

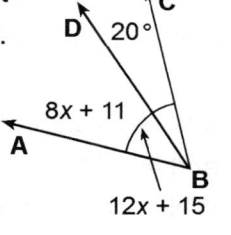

A 63°

B 4°

C 43°

D 51°

16 _____

17 Write "60 pounds for $6.60" as a unit rate per ounce to the nearest tenth.

A $\frac{1 \text{ pound}}{176 \text{ cents}}$

B $\frac{3 \text{ pound}}{33 \text{ cents}}$

C $\frac{1 \text{ pound}}{3.8 \text{ cents}}$

D $\frac{1 \text{ pound}}{0.7 \text{ cents}}$

17 _____

Test 2 – Part 1

18 If the check at a restaurant totals $175.95 and the customer leaves a 20% tip, what is the total cost of the meal, including tip?

A $179.47
B $211.14
C $217.90
D $197.15   18 _____

19 There are 500 residents in the town of Piney Flats. Each one received a survey to determine if they would favor hosting tryouts for the Great American Talent Quest television show. 50 surveys were returned, and 40 of those returned favored the tryouts. Which one of the following statements is true about the results of the survey?

A The results are invalid since only 50 of the 500 residents (10%) responded.
B The results are valid since 80% of the residents indicated they favored the tryouts.
C The results are valid since all 500 residents received the survey even if they did not respond.
D The results are invalid since the 500 residents were not randomly selected.   19 _____

20 Brandy is icing a cake. She needs to know how much icing to make to completely cover all surfaces except for the bottom of the cake. The cake is 12 inches by 7 inches and 3 inches tall. What is the surface area of the cake to be frosted?

A 141 in.^{2}
B 282 in.^{2}
C 198 in.^{2}
D 252 in.^{2} 20 _____

21 What is the product of $-1\frac{4}{5}$ and $-2\frac{1}{12}$?

A $-\frac{15}{4}$
B $\frac{4}{15}$
C $-\frac{4}{15}$
D $\frac{15}{4}$ 21 _____

22 The bar graph shows a student's scores on five tests. What is the range of these scores?

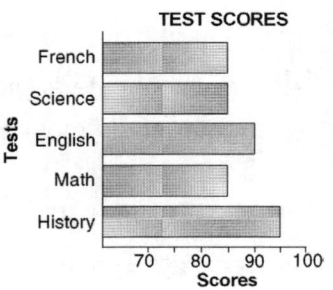

A 10
B 88
C 85
D 15   22 _____

23 Lori bought a turtle for $8.85, two goldfish for $1.25 each, and a snail for $2.95. After she was charged 7% sales tax, how much was her total purchase?

- A $13.96
- B $14.30
- C $15.30
- D $10.01   23 _____

24 Determine which number property is illustrated by the given statement:

$$5 + (-5) = 0$$

- A Commutative Property of Addition
- B Property of Additive Inverse
- C Addition Property of Equality
- D Identity Property for Addition   24 _____

25 A bag holds three blue marbles and four red marbles. What is the probability that a green marble will be picked at random?

- A 1
- B $\frac{1}{4}$
- C 0
- D $\frac{1}{3}$ 25 _____

26 Kenny walks 9 dogs in his dog walking job. He has $42\frac{3}{4}$ feet of rope from which to make leashes of equal length. How long is each length of dog leash?

- A 4.5 feet
- B 4.75 feet
- C 3.75 feet
- D 4.25 feet   26 _____

27 Kara is conducting an experiment to see how many tails land face up when tossing two pennies at the same time. Her results are shown in the chart below.

Tails Showing	Frequency
0	12
1	30
2	8

Based on Kara's results, what is the probability that two tails will land face up when the two pennies are tossed?

- A $\frac{1}{3}$
- B $\frac{8}{50}$
- C $\frac{1}{4}$
- D $\frac{17}{50}$ 27 _____

Test 2 – Part 1

28 Look at the spinner below. What is the probability of the spinner shown landing on a letter used in the word "SPUR"?

A 1
B $\frac{2}{5}$
C $\frac{3}{5}$
D $\frac{1}{5}$

28 _____

29 The circumference of a circle is 20π. What is the area of the circle in terms of π?

A 20π
B 400π
C 10π
D 100π

29 _____

30 The following chart shows how many of each dessert has been sold at a bake sale.

Bake Sale

Dessert	Number Sold
Brownies	64
Cookies	76
Ice cream	38
Pie	28

Based on this chart, what is the probability that the next item sold will be an ice cream?

A about 18%
B about 10%
C exactly 38%
D about 25%

30 _____

31 Max is conducting an experiment in math class with the spinner below.

Event *A* is spinning a prime number while event *B* is spinning an even number. He needs to display the possible outcomes in a Venn diagram. Which number(s) should he write in the intersection of the two circles for event *A* and event *B*?

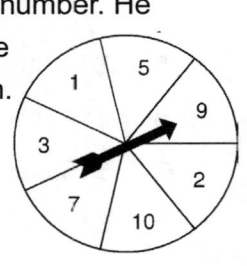

A no number
B 1 and 2
C 1, only
D 2, only

31 _____

32 Charlie rolls a six-sided die and flips a coin. Find the probability that Charlie's die and coin will land in the manner described:

even number and a head

A $\frac{1}{4}$
B $\frac{1}{2}$
C $\frac{3}{8}$
D $\frac{2}{3}$

32 _____

33 Which of the following equations represents the proportion "16 pages written in 3 days is equal to x pages written in one day?"

A $\dfrac{16}{3} = \dfrac{x}{1}$

B $x = 3 + 16$

C $x = 3(16)$

D $\dfrac{16}{3} = \dfrac{1}{x}$

33 _____

34 What is "$60 spent in 5 hours" written as a unit rate in minutes?

A $0.20/min

B $0.40/min

C $0.02/min

D $12/min

34 _____

35 Which one of the following equations does not represent a proportional relationship?

A $\dfrac{1}{2}y = x$

B $5x = y$

C $y = 7x - 3$

D $y = 8x$

35 _____

36 If Hudson Real Estate company is entitled to $\dfrac{5}{8}$ of the commission on the sale of a home and the listing agent receives $\dfrac{2}{5}$ of this, how much does the listing agent receive if the commission totals $4,800?

A $1,200

B $960

C $1,920

D $2,880

36 _____

37 What is the area of the circle pictured below? [Use π = 3.14.]

A 7.065

B 18.84

C 113.04

D 28.26

37 _____

38 Which one of these four graphs shows a proportional relationship between quantities?

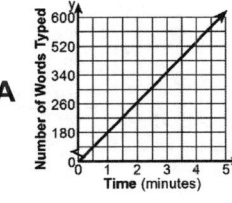
A

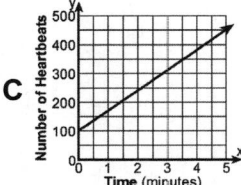
C

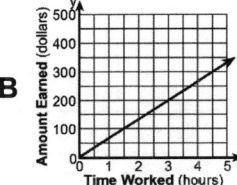
B

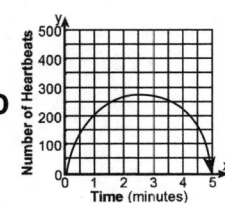
D

38 _____

39 Find the correct completely factored form of the given expression:

$$27x - 18$$

A $-9(3x - 2)$
B $9(3x - 2)$
C $-9(3x + 2)$
D $9(3x + 2)$ 39 _____

40 Hannah went up 340 feet while hiking. If Hannah started at 100 feet below sea level, which integer represents her elevation now?

A +440
B −440
C −240
D +240   40 _____

41 Which one of the following graphs represents the inequality $x < 2$?

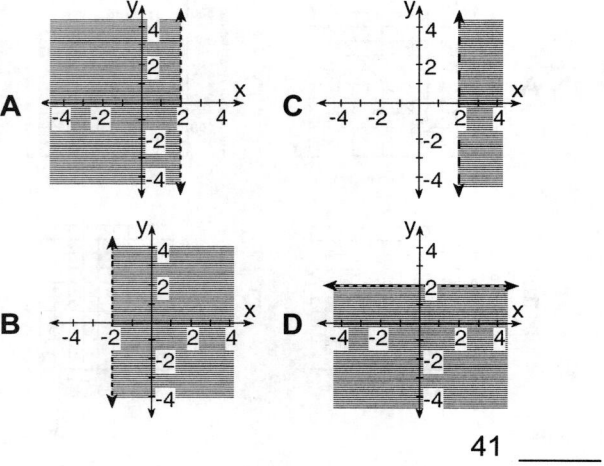

41 _____

42 Which of the following is a valid proportion?

A $\dfrac{2}{13} = \dfrac{6}{72}$

B $\dfrac{1}{20} = \dfrac{80}{1{,}600}$

C $\dfrac{5}{9} = \dfrac{44}{99}$

D $\dfrac{22}{30} = \dfrac{38}{80}$ 42 _____

43 A company budgets $\dfrac{1}{15}$ of its income each month for advertising. In July, the company had an income of $225,000. What is the amount budgeted for advertising in July?

A $10,000
B $5,000
C $20,000
D $15,000   43 _____

44 A number multiplied by 78 is 6,630. What is the number?

A 85
B 90
C 88
D 89   44 _____

45 What is the range of the data in the stem-and-leaf plot below?

Stem	Leaf
2	4 5 6 6 7
3	1 2 5 6 8
4	7 8 9 9 9

KEY: 2|4 means 24

- **A** 15
- **B** 49
- **C** 24
- **D** 25

45 _____

46 Find the correct completely factored form of the given expression:

$$30t - 12$$

- **A** $30(t - 12)$
- **B** $12(2t - 1)$
- **C** $6(5t - 2)$
- **D** $2(15t - 12)$

46 _____

47 On a 15-hour trip, Rob drove 35% of the time and Mark drove the remainder of the time. How many hours did Mark drive?"

- **A** $5\frac{1}{4}$
- **B** $8\frac{1}{2}$
- **C** $9\frac{3}{4}$
- **D** $6\frac{1}{2}$

47 _____

48 The probability of an event happening is 1. This statement means that the event is likely to happen

- **A** never
- **B** always
- **C** half of the time
- **D** only once

48 _____

49 The length of a diameter of a circle is $\frac{2}{a}$. What is the length of a radius of the circle?

- **A** 2
- **B** $\frac{1}{2a}$
- **C** $\frac{1}{a}$
- **D** a

49 _____

50 The circumference of a circle is 35π. What is the radius of the circle?

- **A** 8
- **B** 17.5
- **C** 35
- **D** 70

50 _____

Test 2 – Part 1

51 The probability that it will snow today is 0. What is the probability that it will not snow today?

A $\frac{1}{2}$
B 0
C 100
D 1

51 _____

52 Elizabeth's dad is helping her conduct an experiment for homework. Elizabeth randomly chooses one card from the 10 cards held by her dad, looks at the number, and then replaces the card.

If Elizabeth repeats the process described 100 times, how many times would you expect her to pick the number 3 or the number 5?

A 20
B 25
C 10
D 2

52 _____

53 The total of eight times a number and fifteen is thirtyone. Find the number.

A 2
B 4
C $5\frac{3}{4}$
D −2

53 _____

54 At the beginning of the month, Kiki had $300 in her debit account. She made debit purchases for $125.15 and $250 and she made one deposit of $175.75. Which one of the following best represents her debit account balance after these transactions?

A −$101.50
B −$250.90
C $100.60
D $75.15

54 _____

55 Find the mean of the following numbers:

43, 58, 42, 43, 52, 56

A 43
B 52
C 47
D 49

55 _____

PART 2

56 Betty is making brownies for a bake sale. To the right is a table listing two of the ingredients needed for each batch of 16 brownies.

Servings per Batch	Cups of Cocoa (x)	Cups of Flour (y)
16	1	1.5
32	2	3
48	3	4.5
64	4	6
80	5	7.5

Graph Betty's data on the graph paper below. Be sure to title the graph and label the axes.

Explain why the data you graphed above shows a proportional relationship.

_____
_____
_____
_____
_____

57 Part A

Ms. Mitchell writes this inequality on the board: 2x ≥ 10. Joey solves the inequality and gets the answer x ≤ 5. Is his answer correct? *Explain why or why not.*

_____
_____
_____
_____

Part B

Graph the correct answer on the number line below.

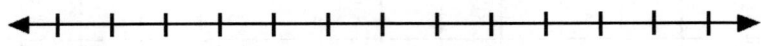

58 Danielle's bird feeder holds $\frac{3}{4}$ cup of birdseed. She is filling the bird feeder with a scoop that holds $\frac{1}{8}$ of a cup. Write a comparison of cups of birdfeed to scoops of birdfeed as a ratio in simplest terms.

Show your work:

*Answer:*_____

59 Find the quotient of $\left(-\frac{4}{3}\right)$ divided by the product of (-2) and $\left(-1\frac{5}{8}\right)$.

*Answer:*_____

60 Emory works 6-hour days selling beverages at an amusement park and earns $10 an hour at her job.

Part A

Write an inequality representing the minimum number of hours (h) Emory will need to work to earn the $5,400 cost of a used car and the 8% state sales tax charge. Show your work.

Inequality: _____

Answer: _____ hours

Part B

On the number line below, graph the solution for the inequality you wrote in Part A represented as the minimum number of full days (d) worked. Show your work.

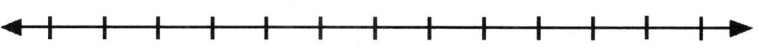

61 A land developer purchased $8\frac{5}{6}$ acres of land for a building project. Two and one-half acres were set aside for a park. How many $\frac{1}{3}$ acre parcels of developed land can be sold?

Show your work:

Answer: _____ parcels

62 The Minnesota state government is concerned that the northeastern Minnesota moose population is declining. In 2012, researchers captured 76 moose, fitted them with radio collars, and released them back into the wild. Several weeks later 168 moose were captured, 3 of which wore radio collars.

Part A
What is the best estimate for the 2012 moose population in northeastern Minnesota?
Show your work:

*Answer:*_____ moose

Part B
If the moose population in 2011 was estimated to be 4,900 moose, what percent decline is shown in this population to the nearest percent?
Show your work:

*Answer:*_____ %

63 7 of the students on Evan's school bus know how to ski and 5 students know how to snowboard. 3 students know both how to ski and how to snowboard. If there are 14 students on the school bus, how many students neither know how to ski nor how to snowboard? Use the Venn diagram below to support your answer.

*Answer:*_____ students

64 Given ∠OMN = 50°, what is the measure of ∠LMN.

Show your work:

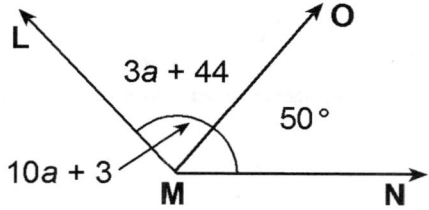

Answer:_____

65 The accompanying table of values shows Callie's pay versus the number of hours she babysat.

Hours Spent Babysitting (h)	Wage (w)
2	$24
5	$60
6	$72
9	$108

Part A

Write an equation to represent the wage in relation to the number of hours worked.

Answer:_____

Part B

How would this equation change if she is given a $5 tip each time?

Answer:_____

Test 3

PART 1

1 What percent of the figure is represented by the shaded part of the diagram below?

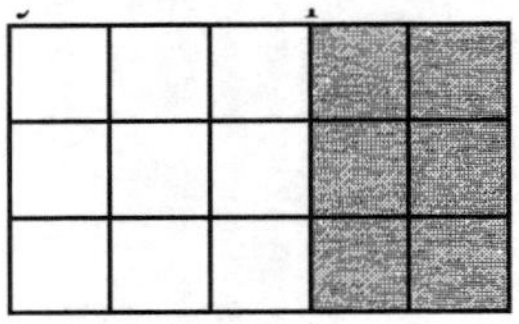

A   60%
B   34%
C   40%
D   25%

1_____

2 What inequality is represented by the graph below?

A $x < -1$
B $x \leq -1$
C $x \geq -1$
D $x > -1$

2_____

3 Write "$13.76 for 4.3 pounds" as a unit rate in ounces.

A $0.86/ounce
B $0.20/ounce
C $3.20/ounce
D $0.27/ounce

2_____

4 Which of the following is a valid proportion?

A $\dfrac{11}{20} = \dfrac{16}{30}$

B $\dfrac{4}{9} = \dfrac{16}{33}$

C $\dfrac{15}{3} = \dfrac{60}{9}$

D $\dfrac{5}{11} = \dfrac{20}{44}$

4_____

5 Simplify the given expression:

$$-3ab + 7ab =$$

A $4a^2b^2$
B $4ab$
C $10ab$
D $-4ab$

5_____

6 Which one of the following tables below shows a proportional relationship between x and y?

A

x	y
2	4
3	9
4	16

B

x	y
2	4
3	5
4	6

C

x	y
2	4
3	6
4	8

D

x	y
2	4
3	3
4	2

6_____

7 The table below shows the amount of calories in a box of chocolates that contains 8 servings.

Calories	220	275	330	385	440
Serving Size (pieces)	4	5	6	7	8

Is the relationship between calories and serving size proportional or non-proportional?

A Proportional because the unit rate is the same.

B Proportional because the unit rate is the different.

C Non-proportional because the unit rate is the same.

D Non-proportional because the unit rate is different.   7 _____

8 Alex wants to buy a portable audio player that sells for $89. He sees it advertised for 30% off? How much should he expect to pay for it?

A $62.30
B $72.48
C $59.33
D $57.00   8 _____

9 James is playing football. On his first play, he gains 10 yards. On the second play, he makes an error and receives a 10-yard penalty. What is his total yardage gain?

A 10 yards
B −10 yards
C 20 yards
D 0 yards   9 _____

10 Add: −19 + (+42)

A −61
B −23
C +23
D +61   10 _____

11 A student read $\frac{3}{5}$ of a book containing 380 pages. How many pages remain to be read?

A 216
B 148
C 152
D 228   11 _____

12 How is the given expression rewritten using the Distributive Property?

$$y + 3y$$

A $y(1 + 3)$
B $3y(y)$
C $3(y)$
D $(-1 + -3)$ 12 _____

13 Nani is presenting plans to ETSU for a new yoga studio she wants to build. The plans show the relative length as 5 inches and the width as 3 inches. If the actual width of the room will be 15 feet, what should be the actual length of the room?

A 25 feet
B 15 feet
C 9 feet
D 5 feet

13 _____

14 What geometric shapes can be drawn from two sets of equivalent parallel sides measuring 3 inches and 4.5 inches?

A A trapezoid and a parallelogram
B A rectangle, a parallelogram, and a square
C A trapezoid, a parallelogram, and a rectangle
D A rectangle and a parallelogram

14 _____

15 What is the supplement of an angle that measures $(3x)°$?

A $(3x - 180)°$
B $(180 - 3x)°$
C $(90 - 3x)°$
D $(3x - 90)°$

15 _____

16 The Worldwide Moving Company questioned 400 of its employees to obtain the information shown in the chart below.

After the Workday is Done

See two or more movies a month	104
Volunteer work	164
Go to theater, ballet, art galleries once a month	232
Exercise at least once a week	336

The executives want to estimate how many of the company's 1,200 employees would use a new exercise facility. Based on the information given, how many do you predict would use the exercise facility?

A 480 people
B 1,008 people
C 804 people
D 336 people

16 _____

17 Write "150 miles on 12 gallons" as a unit rate per pint in simplest form.

1 gallon = 4 quarts
1 quart = 2 pints

A $\dfrac{25 \text{ miles}}{96 \text{ pints}}$

B $\dfrac{3 \text{ miles}}{2 \text{ pints}}$

C $\dfrac{25 \text{ miles}}{8 \text{ pints}}$

D $\dfrac{25 \text{ miles}}{16 \text{ pints}}$

17 _____

18 A student received grades of 92, 74, 69, 87, and 83 on five English exams. Find the average grade of the student's English exams.

    A   81
    B   76
    C   79
    D   91   18 _____

19 What is the area of a square whose sides each measure 5 feet long?

    A   10 ft²
    B   20 ft²
    C   25 ft²
    D   5 ft²   19 _____

20 In order to gather unbiased data, a sample should be

    A   half the size of the population
    B   representative of the general population
    C   less than ten percent of the general population
    D   representative of the adults in the population   20 _____

21 Angie's daily activities are represented in the diagram.

Which of the following statements can not be assumed about the information in the diagram?

    A   Angie spends 6 hours a day sleeping.
    B   Angie spends around 1 month each year driving.
    C   Angie drives 30 miles to work each day.
    D   Angie spends more time working than on any other activity.   21 _____

22 The spinners below have each been divided into equal sections.

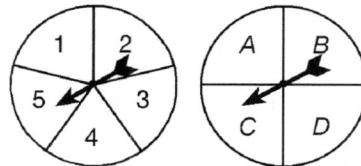

If the two spinners shown are spun at the same time, what is the probability that the spinners will land on an odd number and a vowel?

    A $\frac{3}{9}$
    B $\frac{3}{20}$
    C $\frac{4}{20}$
    D $\frac{4}{9}$ 22 _____

23 Elizabeth's dad is helping her conduct an experiment for homework. Elizabeth randomly chooses one card from the 10 cards held by her dad, looks at the number, and then replaces the card.

If Elizabeth repeats the process described 150 times, how many times would you expect her to pick a multiple of 3?

A 50
B 100
C 30
D 45   23 _____

24 Find the median of the following numbers:
14, 27, 13, 33, 27, 9, 11, 26

A 20
B 27
C 14
D 26   24 _____

25 From a $13\frac{1}{2}$-inch long piece of ribbon, how many $4\frac{1}{2}$-inch lengths can be cut?

A 3
B 4
C 2
D 5   25 _____

26 Simplify the given expression: $7x^2y - 3x^2y =$

A $10x^2y$
B 4
C $-4x^2y$
D $4x^2y$ 26 _____

27 Which one of the following graphs represents the solution of the inequality $2x + 3 > 9$?

A (number line -1 to 5, open circle at 3, shaded left)
B (number line -1 to 5, closed circle at 3, shaded left)
C (number line -1 to 5, closed circle at 3, shaded right)
D (number line -1 to 5, open circle at 3, shaded right)

27 _____

28 Solve the equation for the given variable:
$3z - 19 = -31$

A -7
B -4
C -3
D -16 28 _____

29 $6z + 8 - 5z = 83$

- A 39
- B 75
- C 91
- D 37

29 _____

30 What is the probability that it will snow if the temperature is 75°F?

- A 0.5
- B 0.75
- C 0
- D 1

30 _____

31 If a standard 6-sided die is tossed once, the probability of getting a 5 is $\frac{1}{6}$. What is the probability of not getting a 5?

- A $\frac{5}{6}$
- B $\frac{1}{6}$
- C $\frac{1}{3}$
- D $\frac{4}{5}$

31 _____

32 Which of the following is a valid proportion?

- A $\frac{3}{5} = \frac{34}{60}$
- B $\frac{3}{8} = \frac{24}{50}$
- C $\frac{42}{48} = \frac{7}{8}$
- D $\frac{4}{7} = \frac{18}{35}$

32 _____

33 A home has a mortgage of $64,000 for 30 years at an annual interest rate of 15%. Find the monthly mortgage payment rounded to the nearest cent.

- A $860.65
- B $809.24
- C $820.37
- D $802.94

33 _____

34 The spinners shown below are each spun once.

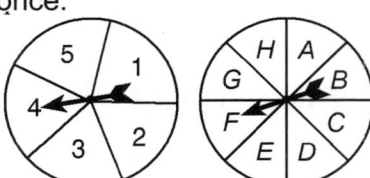

What is the probability of spinning an even number and a vowel?

- A $\frac{1}{20}$
- B $\frac{1}{40}$
- C $\frac{6}{13}$
- D $\frac{1}{10}$

34 _____

Test 3 – Part 1

35 Which of the following histograms accurately shows the number of hours the people surveyed sleep?

Number of Hours of Sleep	Frequency
4-5	6
6-7	17
8-9	13
10-11	5

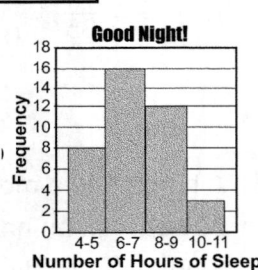

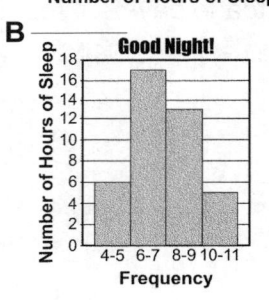

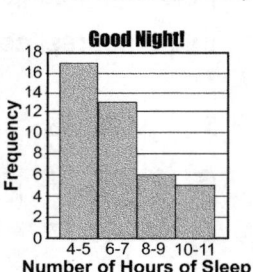

35 _____

36 0.085% of 1,200 is what?

A 10.2
B 1.02
C 113.6
D 0.7

36 _____

37 The accompanying table shows the number of milligrams of active ingredients of a cold medication remaining in the bloodstream x hours after consumption.

Approximately how many hours will it take until no active ingredients remain in the bloodstream?

x (hours)	y (mg)
1	122
2	104
5	50

A 7 hours
B 6 hours
C 9 hours
D 8 hours

37 _____

38 Eva is conducting an experiment with a standard 6-sided die. Event A is rolling a number less than 3 while event B is rolling a prime number. She wants to display the possible outcomes in a Venn diagram. Which number(s) should she write in the intersection of the two circles for event A and event B?

A 2, only
B 2 and 3
C 1 and 2
D 1, 2, and 3

38 _____

39 Which of the following is a valid proportion?

A $\dfrac{2}{3} = \dfrac{66}{90}$

B $\dfrac{3}{5} = \dfrac{30}{70}$

C $\dfrac{3}{4} = \dfrac{45}{60}$

D $\dfrac{4}{5} = \dfrac{40}{60}$

39 _____

40 Belinda buys a pair of sandals for *s* dollars. She must pay a sales tax of 7% on her purchase. What expression represents how many dollars she will pay altogether, including the sales tax?

A   7s
B   s + 0.07
C   1.07s
D   1.7s   40 _____

41 A sales representative for a hardware company receives $8,000 per year plus a 6% commission on total sales. During one year the sales representative's sales totaled $320,000. Find the representative's total earnings for the year.

A $27,200
B $19,200
C $22,700
D $32,200   41 _____

42 The sales tax on a radio is $2.45. If the cost of the radio is $35, what is the sales tax rate?

A   8%
B   8.25%
C   6%
D   7%   42 _____

43 Which one of the following situations represents a relationship equal to 0?

A   running 4 km west and walking 4,000 m south
B   gaining 6 pounds and losing 72 ounces
C   heating an ice cube to 80°F then freezing it at 0°C
D   earning $12 and spending 120¢
                            43 _____

44 In a living room, a square coffee table is 6 feet long. In a scale model of the table, the scale is 1 inch = 3 feet. What should be the length of the model?

A   4 inches
B   6 inches
C   2 inches
D   3 inches   44 _____

45 The diagram below shows a right circular cone with four possible cutting paths, *A*, *B*, *C*, and *D*.

What letter in the given diagram represents the path that, when cut, would result in a triangular cross-section?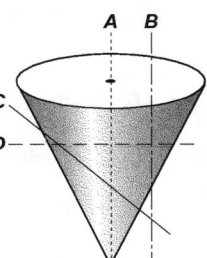

A   *A* and *C*, only
B   *A*, only
C   *A* and *B*, only
D   *A*, *B*, and *C*, only   45 _____

46 What is the greatest common factor of 42 and 56?

A 2
B 14
C 7
D 8   46 _____

47 The following spinner is divided into 6 equal sections.

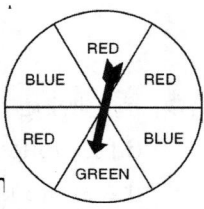

Predict how many times out of 480 spins the spinner shown will most likely stop on the color blue.

A 108
B 80
C 240
D 160   47 _____

48 If two angles of a triangle measure 48° and 42°, the triangle is

A right.
B isosceles.
C acute.
D equilateral.   48 _____

49 The diagram represents a cube that has been cut along a path parallel to face A.

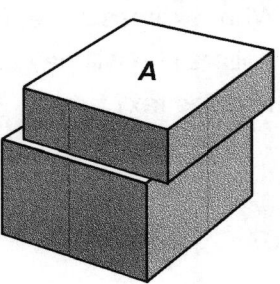

What is the shape of the two-dmensional surface that results from this cross-section?

A square
B right triangle
C trapezoid
D hexagon   49 _____

50 Given ∠JKL = 120°, what is the measure of ∠JKM.

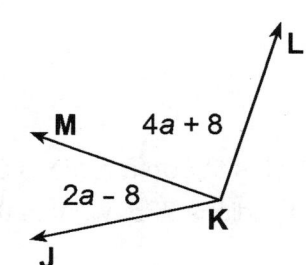

A 40°
B 20°
C 88°
D 32°   50 _____

Page 38 — Test 3 – Part 1 — Copyright © 2016 Topical Review Book Company

51 A rectangular pool sits inside a grassy rectangular yard measuring 104 feet by 95 feet. The dimensions of the pool are 19 feet by 43 feet.

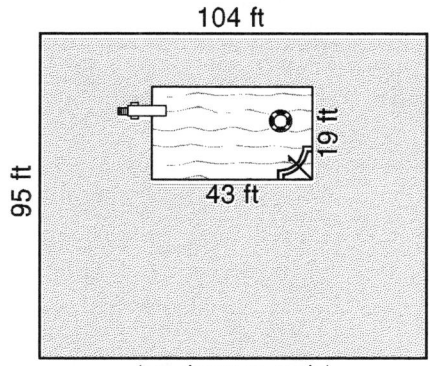

(not drawn to scale)

What is the approximate size of the grassy area of the yard (in square feet)?

A 10,800 ft²
B 8,000 ft²
C 9,200 ft²
D 10,000 ft²

51 _____

52 Which of the following is a valid proportion?

A $\frac{5}{18} = \frac{14}{60}$
B $\frac{7}{12} = \frac{84}{144}$
C $\frac{7}{9} = \frac{44}{63}$
D $\frac{8}{15} = \frac{94}{160}$

52 _____

53 Kortney is planning a trip for her History club. Which of the following would provide the most representative survey sample to determine the preferred location for their trip?

A all the History club leaders
B all female club members of the History club
C all the History teachers in the school
D all members of the History club

53 _____

54 A company spends $\frac{3}{8}$ of its monthly income on employee salaries. During the month of May, the company had an income of $262,400. How much of the May income remains after the employee salaries are paid?

A $96,000
B $98,400
C $164,000
D $160,000

54 _____

55 Find the product of the given set of numbers:

(−12)(−7)

A 84
B −84
C 96
D −96

55 _____

Test 3 – Part 1

56 A toothbrush factory has to reach a daily minimum quota of 420 toothbrushes made every day. Unfortunately this morning at 10 AM a machine broke. If they already made 160 toothbrushes and continue to make 35 toothbrushes every hour thereafter, can they meet their quota before the end of the workday at 5 PM?

Show your work:

*Answer:*_____

Explain your answer using words.

_____
_____
_____
_____

57 A company selling hospital equipment pays its sales executives a commission of 8% of all additional sales over $200,000. During one year a sales executive sold $1,450,000 worth of hospital equipment. Find the commission earned by the sales executive.

Show your work:

*Answer:*_____

58 A community garden in the shape of a circle has a circumference of 30π yards. What is the area of the garden in terms of π?

Show your work:

Answer:_____ yd²

59 The pictograph below shows three specific types (genre) of used CDs sold.

What is the range in the number of used CD sales represented by the pictograph shown?

Show your work:

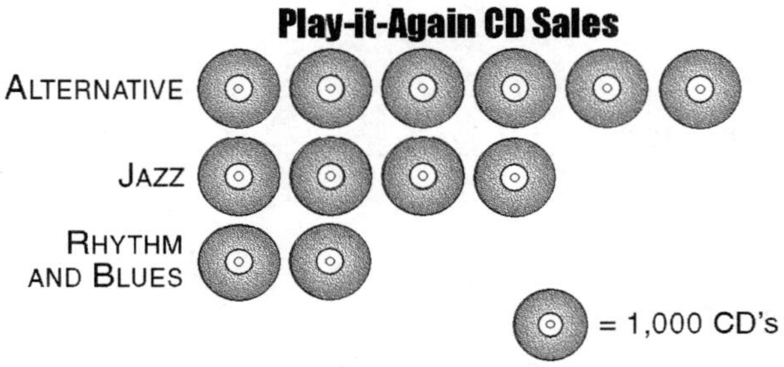

Play-it-Again CD Sales

ALTERNATIVE ◉ ◉ ◉ ◉ ◉ ◉
JAZZ ◉ ◉ ◉ ◉
RHYTHM AND BLUES ◉ ◉

◉ = 1,000 CD's

Answer:_____

60 Describe an event that is as likely to happen as not.

_____
_____
_____
_____
_____

61 Neil is on a trip to Australia. He records his cell phone usage so he can track the money he spends. The information is listed in the accompany table.

Length of Call (minutes)	Cost (dollars)
1	0.25
2	0.50
3	0.75
4	1.00
5	1.25
6	1.50

Part A

Write an equation that represents the relationship between the data shown. Use m for the length of a call in minutes and d for the cost in dollars.

Answer:_____

Part B

Use your equation from Part A to determine how much an 18-minute call will cost in dollars.

Show your work:

Answer:_____

62 Rinspeed's new sQuba is the world's first diving car able to both drive on roads and beneath the water. During a test drive, this car traveled 6.2 miles on dry land from an elevation of 244 yards, submerged below sea level to 11 yards for 2 miles across a channel to the other side, and then back up onto dry land for 2.9 miles to an elevation of 53 yards. If the car travels an average of 24,640 yards per gallon of gasoline, about how much gasoline (to the nearest tenth) did this car use to complete both the vertical and horizontal components of this trip?

1 mile = 1,760 yards

Show your work:

Answer:_____gal

63 Each section in the spinner below divides the circle into section in the following measures: 35% red, 15% green, 28% teal, and 22% purple.

Do you think the results from this table are possible outcomes from an experiment conducting 100 spins of the accompanying spinner?

Justify your reasoning.

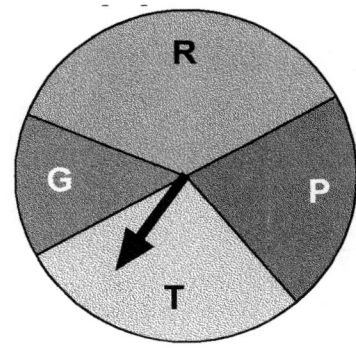

Color	Frequency
Red (R)	|||| |||| |||| |||| ||||
Green (G)	|||| |||| |||| |||| |||| ||||
Teal (T)	|||| |||| ||
Purple (P)	|||| |||| |||| |||| |||| |||| |||

64 Four friends decided to wash cars one month during the summer. Jeffrey washed 45 cars that month. Anna washed 60 cars, Genevieve washed 25 cars, and Troy washed 70 cars that month. If they raised a total of $1,450 altogether and each of them plan to take home the same fraction of money as compared to the fraction of cars they washed, how much money will Anna have earned?

Show your work:

Answer: $_____

65 Belinda has a gift card for $50 towards the purchase of apps for her phone. The apps she wants to purchase cost $2.75 each.

Part A

Write and solve an inequality that shows the maximum number of apps (*a*) Belinda can purchase with a $50 gift card.

Show your work:

*Inequality:*_____

*Answer:*_____ apps

Part B

Graph the solution set for the inequality you wrote in Part *A* on the number line below. Be sure you show that you can not purchase fewer than zero apps.

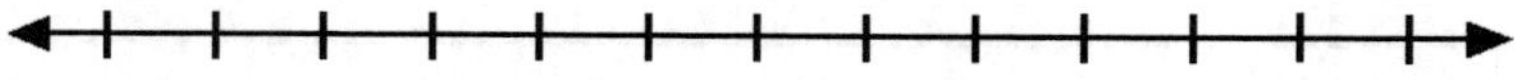

Test 4

PART 1

1 Ellen deposited $2,500 into a savings account that earns 5% interest per year. Her friend's bank offers a $6\frac{1}{2}$% annual interest rate. How much more money would Ellen's money have earned in one year if she had deposited her money at her friend's bank?

A $12.50
B $32.50
C $37.50
D $16.25   1 _____

2 Solve the equation for the given variable:
$$-3x - 4(2 - x) = -5$$

A $-\frac{3}{4}$
B 3
C $-1\frac{1}{3}$
D −13   2 _____

3 What is "180 square feet of wall painted in 12 minutes" written as a unit rate in seconds?

A 3 square feet of wall painted/second
B 0.25 square feet of wall painted/second
C 900 square feet of wall painted/second
D 15 square feet of wall painted/second

3 _____

4 Which one of the following charts shows a proportional relationship between quantities?

A
Number of Games	1	3	5	7	9
Number of Goals	6	15	24	33	42

B
Number of Games	1	3	5	7	9
Number of Goals	5	10	15	20	25

C
Number of Games	1	3	5	7	9
Number of Goals	8	16	24	32	40

D
Number of Games	1	3	5	7	9
Number of Goals	4	12	20	28	36

4 _____

5 Simplify the given expression:
$$24xy - xy =$$

A 24
B $-23xy$
C $23xy$
D $25xy$ 5 _____

Test 4 – Part 1

6 The Somber Mask Drama Club is putting on a pancake breakfast as a fundraiser. The number of pancakes depends on the number of cups of batter used.

What quantity of batter is used for each pancake?

Number of Cups	Number of Pancakes
1	4
3	12
4	16
6	24

A $\frac{1}{3}$ cup

B $\frac{1}{2}$ cup

C 1 cup

D $\frac{1}{4}$ cup

6 _____

7 If twelve pounds of nails cost $3.84, what is the unit cost?

A 32¢
B 28¢
C 23¢
D 18¢

7 _____

8 A department store uses a markup rate of 35% on all its calculators. What is the selling price of a calculator which cost the store $62?

A $83.70
B $87.30
C $93.70
D $21.70

8 _____

9 Find the sum of the given set of numbers:
$$(-27) + (-29) + 50$$

A 6
B –6
C –48
D 48

9 _____

10 Solve the given expression:
$$-5\frac{1}{4} + -8\frac{1}{5} =$$

A $-3\frac{1}{20}$

B $\frac{9}{20}$

C $-\frac{9}{20}$

D $3\frac{1}{20}$

10 _____

11 Jackie bought three pounds of fish. Her recipe calls for $\frac{1}{4}$ pound of fish per serving. How many people will the fish feed?

A 12
B 6
C 9
D 15

11 _____

12 How is the given expression rewritten using the Distributive Property?

$$7x + (-2x)$$

A $-x(7 + 2)$

B $x(7 + 2)$

C $-7(x + -2)$

D $-2(x + -7)$ 12 _____

13 The distance between Miami, FL and Washington DC on a map is 12.5 centimeters. If the map scale is 2 centimeters = 160 miles, what is the actual distance between the two locations?

A   320 miles

B   1,000 miles

C   2,000 miles

D   2,500 miles   13 _____

14 If the sum of the measures of two angles of a triangle is equal to the measure of the third angle, the triangle must be

A   obtuse.

B   acute.

C   right.

D   isosceles.   14 _____

15 Which of the following is a valid proportion?

A $\dfrac{3}{2} = \dfrac{40}{65}$

B $\dfrac{2}{9} = \dfrac{5}{18}$

C $\dfrac{17}{51} = \dfrac{12}{8}$

D $\dfrac{3}{2} = \dfrac{90}{60}$ 15 _____

16 If the actual distance between two locations is 10 miles and the map scale is 2 inches = 5 miles, what is the distance between the two locations on the map?

A   10 inches

B   20 inches

C   8 inches

D   4 inches   16 _____

17 Write "192 miles on 12 gallons" as a unit rate per quart in simplest form.

$$\boxed{1\text{ gallon} = 4 \text{ quarts}}$$

A $\dfrac{64 \text{ miles}}{1 \text{ quart}}$

B $\dfrac{1 \text{ quart}}{4 \text{ miles}}$

C $\dfrac{4 \text{ miles}}{1 \text{ quart}}$

D $\dfrac{16 \text{ miles}}{1 \text{ quart}}$ 17 _____

18 The bar graph shows a student's scores on five tests.

What is the mean of these scores?

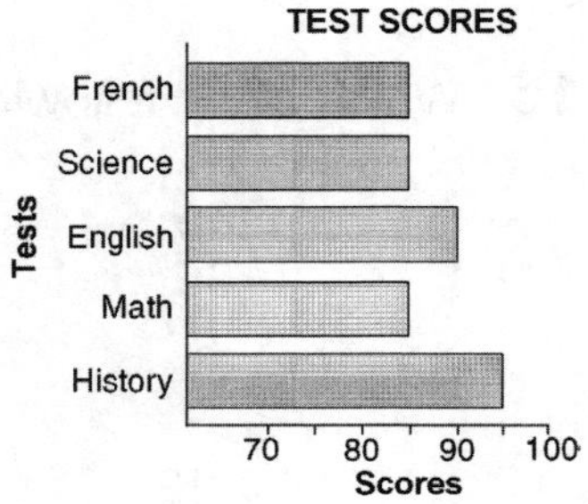

TEST SCORES
(bar graph showing: French ~85, Science ~85, English ~90, Math ~85, History ~95)

A 90
B 88
C 80
D 85   18 _____

19 What would be the cost of refinishing a hardwood floor 11 feet 2 inches by 9 feet 10 inches at $2.50 per square foot?

A $105.00
B $329.40
C $274.40
D $52.50   19 _____

20 Jana is in charge of a fundraiser for Pendelton Elementary School. She wants to conduct a survey to see what items the students want to sell. Which group would give the **best** representative sample?

A 67 sixth-grade students
B 67 students selected randomly from grades 1–6
C 67 students who participate in after-school activities
D 67 students from grades 1–6 not participating in the fundraiser   20 _____

21 Find the mean of the following numbers:

64, 76, 63, 69, 72, 76

A 70
B 71
C 72
D 76   21 _____

22 Eva is conducting an experiment with a standard 6-sided die. Event A is rolling a number less than 3 while event B is rolling a prime number. She wants to display the possible outcomes in a Venn diagram. Which number(s) should she write in the intersection of the two circles for event A and event B?

A 1, 2, and 3
B 2 and 3
C 1 and 2
D 2, only   22 _____

23 The salaries of six executives of a computer company are $87,000, $62,400, $54,600, $66,000, $75,400, and $59,000. Find the median salary of the six executives.

A $66,000
B $67,400
C $62,400
D $64,200   23 _____

24 Camille is playing a game in the car with her brother, Cory. With the backs of the cards facing them both, Cory picks a card at random, notes the suit, and then returns the card to his sister's hand. Camille then has to guess the drawn card's suit.

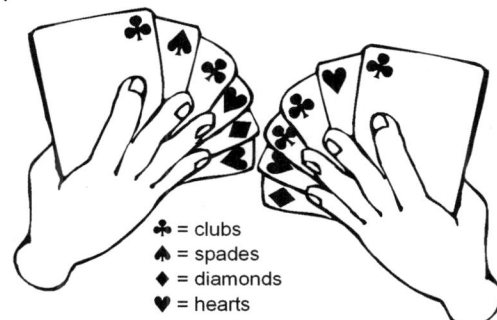

♣ = clubs
♠ = spades
♦ = diamonds
♥ = hearts

If Cory repeats the process described 180 times, predict how many times you would expect him to pick a diamond, spade, or club card.

A 120
B 60
C 135
D 75   24 _____

25 A jeweler purchased a $\frac{5}{8}$ carat diamond for $750. What would be the cost of a similar diamond weighing one carat?

A $1,200
B $1,500
C $1,000
D $468   25 _____

26 Simplify the given expression: $15y^3 - 6y^3 =$

A 9
B $9y^3$
C $9y^6$
D $-9y^3$ 26 _____

27 What inequality is represented by the graph below?

A $-5 \leq x \leq 6$
B $-5 < x < 6$
C $-5 \leq x < 6$
D $-5 < x \leq 6$ 27 _____

28 Solve the equation for the given variable:

$$8 = 5z - 27$$

A 1
B 5
C 3
D 7   28 _____

Test 4 – Part 1

29 What is the probability of a triangle having more than one internal obtuse angle?

A $\frac{1}{2}$

B $\frac{1}{3}$

C 0

D 1

29 _____

30 What is the probability that the sun will not rise tomorrow?

A 100

B 0.5

C 0

D 1

30 _____

31 When a number is chosen at random from the set {1, 2, 3, 4, 5, 6}, which one of the following events has the greatest probability of occurring?

A choosing a number greater than 3

B not choosing either 1 or 6

C choosing a prime number

D choosing an even number

31 _____

32 Which of the following is a valid proportion?

A $\frac{16}{184} = \frac{2}{23}$

B $\frac{5}{18} = \frac{15}{72}$

C $\frac{2}{5} = \frac{10}{35}$

D $\frac{50}{70} = \frac{10}{17}$

32 _____

33 A plumber purchases a used truck for $9,200 and must pay a sales tax of 6% of the purchase price. Find the sales tax.

A $525

B $425

C $552

D $5,648

33 _____

34 Shania is conducting an experiment with a standard 6-sided die. Event A is rolling a number greater than 4 while event B is rolling an even number. She wants to display the possible outcomes in a Venn diagram. Which number(s) should she write in the intersection of the two circles for event A and event B?

A 2, 4, and 6

B 4 and 6

C 2, 4, 5, and 6

D 6 only

34 _____

35 What is the probability of a week consisting of 175 hours?

A 100
B 0.5
C 0
D 1   35 _____

36 92% of 115 is what number?

A 105.80
B 80.00
C 125.00
D 97.34   36 _____

37 Marcy's coed gymnastic team competed in a local competition. One team member performed in each event and their scores are noted below.

Floor Exercise	8.7
Balance Beam	9.3
Men's Vault	8.9
Women's Vault	9.6
Uneven Bars	9.2
Rings	9.4
Parallel Bars	8.9
Pommel Horse	9.0

What is the mode of the team's scores?

A 8.9
B 9.1
C 9.2
D 0.9   37 _____

38 Selma threw a penny, a dime, and a quarter into a fountain. What is the probability all 3 coins will settle tails face up?

A $\frac{1}{2}$
B $\frac{1}{6}$
C $\frac{1}{3}$
D $\frac{1}{8}$ 38 _____

39 When rice is prepared, the amount of rice varies proportionately to the amount of water required. If 2 cups of rice requires 4.5 cups of water, then what is the total number of cups of water needed to prepare 1 cup of rice?

A 2 cups
B 2.25 cups
C 2.5 cups
D 2.75 cups   39 _____

40 Damian buys a boat. He buys gasoline at a unit rate of g dollars per gallon. He puts 24.3 gallons of gasoline in the boat's main tank and puts 0.70 gallons of gasoline in its reserve tank. Which expression represents how many dollars he spent on gasoline?

A $24.3g + 0.07$
B $25g$
C $24.3g + 0.70$
D $24.3 + 0.70$ 40 _____

Test 4 – Part 1

41 A new labor contract called for a 6.5% increase in pay for all employees. What is the new wage of an employee who was making $560 per week?

A $924.00
B $563.64
C $36.40
D $596.40   41 _____

42 Wayne answered 58 out of 65 questions on a test correct. What was the percentage he answered incorrectly?

A 91%
B 89%
C 9%
D 11%   42 _____

43 Simplify: $4 - \dfrac{1}{0.4} =$

A 2.4
B 1.5
C 1.2
D 3.4   43 _____

44 What is the value of y when $\angle LMN = 170°$?

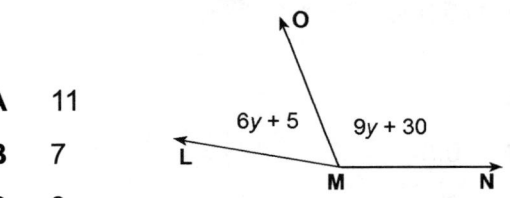

A 11
B 7
C 9
D 14   44 _____

45 The diagram below shows a right circular cone with four possible cutting paths, A, B, C, and D. What letter in the given diagram represents the path that, when cut, would result in an oval cross-section?

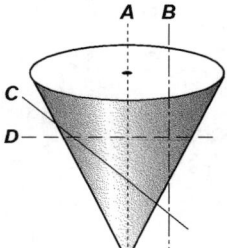

A A and C, only
B C and D, only
C C, only
D D, only   45 _____

46 Find the correct completely factored form of the given expression: $-24g - 2h$

A $-2(12g + h)$
B $-2(-12g + h)$
C $2(12g - h)$
D $-2(-12g - h)$ 46 _____

47 Elizabeth's dad is helping her conduct an experiment for homework. Elizabeth randomly chooses one card from the 10 cards held by her dad, looks at the number, and then replaces the card.

How many times would you expect Elizabeth to repeat the process described in order to pick a number greater than six 60 times?

A 24
B 150
C 600
D 120   47 _____

48 If two angles of a triangle each measure 70°, the triangle is described as

A isosceles.
B equilateral.
C right.
D obtuse.   48 _____

49 If two angles of a triangle measure 43° and 48°, the triangle is

A isosceles.
B acute.
C obtuse.
D right.   49 _____

50 Given ∠ABC = 178°, what is the measure of ∠ABD.

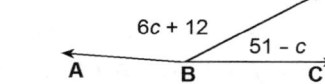

A 126°
B 28°
C 23°
D 150°   50 _____

51 What is the surface area of a cube whose side measures 3 cm?

A 324 cm²
B 36 cm²
C 54 cm²
D 12 cm²   51 _____

52 Which of the following is a valid proportion?

A $\frac{6}{9} = \frac{60}{80}$

B $\frac{7}{35} = \frac{14}{80}$

C $\frac{5}{14} = \frac{30}{72}$

D $\frac{65}{85} = \frac{13}{17}$ 52 _____

Test 4 – Part 1

53 Jada's Adventure troop is going to sell either wrapping paper or gift bags to raise money for their annual trip. The troop will distribute a survey to each member's neighbors to determine which item might sell best. Which one of the following would be the best question to ask on the survey?

    A   How many gifts do you generally give and receive each year?

    B   What do you prefer to use when giving a gift? A gift bag or wrapping paper?

    C   What holidays or occasions do you generally decorate gifts for?

    D   Which item would you like most to sell? Gift bags or wrapping paper?

53 _____

54 A sales executive has completed $\frac{3}{5}$ of an 875-mile business trip. How many miles of the trip remain?

    A   525

    B   495

    C   350

    D   555

54 _____

55 Find the product of the given set of numbers: $(-36)(+\frac{1}{9})$

    A   –4

    B   4

    C   –6

    D   6

55 _____

PART 2

56 Heather is at an Internet cafe. It costs her an initial charge of $1.25 plus an additional $0.10 each minute she uses their wireless connection to browse the Internet.

Part A
Write and solve an inequality that shows the maximum number of minutes Heather can use the wireless Internet connection with $20 to spend. Let x = the number of minutes she can use the wireless connection.

Show your work:

Inequality: _____

Answer: _____ minutes

Part B
On the number line below, graph the solution set for the maximum number of hours Heather can use the wireless connection. Be sure you show that you can *not* browse fewer than zero hours. Show your work.

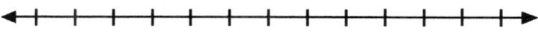

57 Marley is an avid birdwatcher and is trying to estimate the number of Blue Jays living in his region. He traps 22 Blue Jays and tags them. Later, he traps 550 Blue Jays. Based on the number of tags observed in this group, he determines the Blue Jay population to be 825. What is the best estimate for the number of tagged Blue Jays that were recaptured?

Show your work:

Answer: _____ Blue Jays

58 Find the area of a circle with a radius of 14 inches. [Use $\frac{22}{7}$ for π.]

Show your work:

Answer: _____

59 What is the difference between the ranges of the data represented by these two stem-and-leaf plots?

Show your work:

Stem	Leaf
0	4 5 6
1	4 4 7

Stem	Leaf
4	1 5 6
5	0 3 9

KEY: 0|4 means 04 OR 4

Answer: _____

60 Describe a certain event and an impossible event.

_____
_____
_____
_____

61 Below is a recipe card for chocolate chip cookies.

If Ms. Sweet needs to make 15 dozen cookies for the town meeting, how much of each ingredient will she need? Write your answers on the lines below.

Show your work:

_____ c margarine

_____ c sugar

_____ tsp vanilla

_____ c flour

_____ tsp baking powder

_____ tsp salt

_____ oz chocolate chips

COOKIES

1 ½ cups margarine
1 ¾ cups sugar
2 teaspoons vanilla
3 ¼ cups flour
1 teaspoon baking powder
¼ teaspoon salt
8 ounces chocolate chips

<u>Chocolate Chip Cookies</u>

Makes 3 dozen

---

62 A cone-shaped paper water cup was tossed into the air and how the cup landed was recorded in the table below.

Flip of a Cone-shaped Water Cup

S	S	S	T	S	T	T	T
S	S	T	T	S	T	S	S
T	S	T	S	T	S	T	S
S	S	T	T	S	S	S	S
S	T	T	S	S	S	S	S

KEY: S = Side down   T = Top down

Based on this data, is the cup more likely to land on its side or on its top?

Justify your reasoning on the lines below.

_____

_____

_____

_____

63 Patricia went to an amusement park with $60.00. She spent half on her admission ticket and then the $\frac{2}{3}$ of what was left on food and games. How much money did Patricia go home with?

Show your work:

Answer: $_____

64 Kelsey loves to read fairy tale books. She has read 17 books so far. Each month she reads 3 more.

Part A

How many months will it take her to read over 40 books? Write an inequality and solve it.

Show your work:

Inequality: _____

Answer: _____ months

Part B

Graph the solution set for the inequality you wrote in Part *A* on the number line below.

Test 5

PART 1

1 A single card is drawn from a standard deck of 52 cards. What is the probability the card is a five?

A $\frac{13}{52}$

B $\frac{1}{13}$

C $\frac{4}{13}$

D $\frac{1}{4}$ 1 _____

2 Determine which number property is illustrated by the given statement:

$$-3 + 6 = 6 + -3$$

A Property of Additive Inverse

B Associative Property of Addition

C Commutative Property of Addition

D Distributive Property   2 _____

3 Solve the equation for the given variable:

$$5x + 9 = -11$$

A −20

B $\frac{2}{5}$

C −4

D 4   3 _____

4 A computer malfunctioned 19 hours out of a total of 500 hours of computer operation. What percent of the total time of operation was the computer malfunctioning?

A 38%

B 3.8%

C 2.8%

D 26.32%   4 _____

5 What is 21 feet of bricks laid in 7 minutes" written as a unit rate per hour?

A 180 feet of bricks laid/hour

B 420 feet of bricks laid/hour

C 90 feet of bricks laid/hour

D 3 feet of bricks laid/hour   5 _____

6 Simplify the given expression:

$$-2xy + 4xy + 6xz + 3xy =$$

A $11xyz$

B $9xy + 6xz$

C $-5xy + 6xz$

D $5xy + 6xz$ 6 _____

7 Solve the equation for the given variable:

$$-78 = 5y + 27 - 8y$$

A 35
B 26
C 17
D 21

7 _____

8 This month's salary for a vacuum salesman was $2,600. This includes the salesman's base monthly salary of $800 plus a 4% commission on total sales. Find the salesman's total sales for the month.

A $85,000
B $60,000
C $104
D $45,000

8 _____

9 46% of 234 is what number?

A 107.64
B 49.51
C 196.67
D 50.87

9 _____

10 Simplify the given expression:

$$24xy + 13xy =$$

A $37xy$
B $37x^2y^2$
C 37
D $37x^2y$

10 _____

11 Which of the following is a valid proportion?

A $\frac{9}{17} = \frac{32}{67}$
B $\frac{7}{18} = \frac{17}{56}$
C $\frac{3}{11} = \frac{27}{99}$
D $\frac{5}{13} = \frac{23}{82}$

11 _____

12 50% of 70 is what?

A 30
B 40
C 35
D 140

12 _____

13 Write "258.4 miles on 9.5 gallons" as a unit rate in quarts.

A 6.5 miles/quart
B 6.8 miles/quart
C 9.1 miles/quart
D 13.6 miles/quart

13 _____

14 Danielle found that of the 156 cookies she made, 36 were burnt. Approximately what percent of the cookies were burnt?

A 17%
B 41%
C 56%
D 23%

14 _____

15 According to the graph shown, what is the range in the town's population over the years?

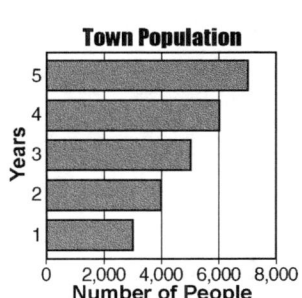

A 7,000 people
B 3,000 people
C 10,000 people
D 4,000 people

15 _____

16 Add: −42 + (+38)

A −70
B −4
C +70
D +80

16 _____

17 The diagram shows a numbered mat at which students toss beanbags.
If a beanbag is tossed at the mat shown, what is the probability that the beanbag will land on a 5?

A $\frac{3}{9}$
B $\frac{5}{9}$
C $\frac{1}{9}$
D $\frac{7}{9}$

17 _____

18 What percent of the figure is represented by the shaded part of the diagram below?

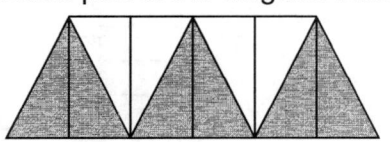

A 57%
B 60%
C 43%
D 40%

18 _____

Test 5 – Part 1

19 Find the product of the given set of numbers:

$$\left(-\frac{3}{5}\right)(-35)$$

- A −15
- B 21
- C 15
- D −21

19 _____

20 Simplify the given expression:

$$3x + 4x =$$

- A $7x^2$
- B $-x$
- C x^7
- D $7x$

20 _____

21 A student is studying actions and reactions. She carries out an experiment to test how high a ball will bounce if she drops it from various heights. The chart below shows the results.

What is the constant of proportionality for the data shown?

Height of Ball When Dropped (in cm)	Height Ball Bounces (in cm)
30	10
60	20
90	30
120	40
150	?

- A 1
- B 5
- C 10
- D 3

21 _____

22 If $\frac{1}{8}$ of the 6,400 students at Morris High School were absent, how many students attended school?

- A 800
- B 5,200
- C 5,600
- D $\frac{7}{8}$

22 _____

23 What is (−13.2) divided by (−12)?

- A 11
- B −11
- C 1.1
- D −1.1

23 _____

24 What is the numerical value of the expression:

$$-2 \cdot (6 - 7) - 5$$

- A −3
- B 3
- C 8
- D −8

24 _____

Test 5 – Part 1

25 In △ABC, m∠A = 41° and m∠B = 48°. What kind of triangle is △ABC?

A   Isosceles
B   Right
C   Acute
D   Obtuse   25 _____

26 The diagram below shows a right circular cylinder with four possible cutting paths, 1, 2, 3, and 4.

What two-dimensional shape will result from a cross-section along path 4 in the given diagram?

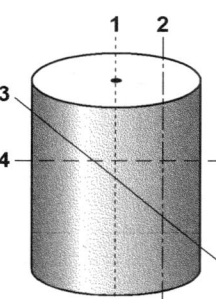

A   parabola
B   rectangle
C   oval
D   circle   26 _____

27 What is the value of x when m∠DEF = 54°?

A   4
B   15
C   12
D   11.5

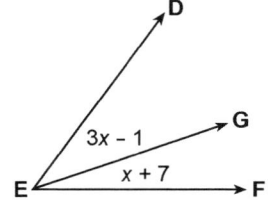
                                          27 _____

28 What is "$96 spent in 4 hours" written as a unit rate in minutes?

A $0.40/minute
B $1.60/minute
C $0.53/minute
D $0.30/minute   28 _____

29 What percent of the figure below is shaded?

A   40%
B   44.4%
C   55.6%
D   50%

                                          29 _____

30 Malcolm is cutting posterboard to make a giant set of dice for a carnival game as shown below.

What is the surface area of one of these die?

A   324 in.²
B   5,832 in.²
C   1,944 in.²
D   108 in.²   30 _____

Test 5 – Part 1   Page 63

31 A school wants to add a coed soccer program. To determine student interest in the program, a survey will be taken. In order to get an unbiased sample, which group should the school survey?

A every third student entering the building

B every member of the varsity football team

C every student having a second-period French class

D every member in Ms. Zimmer's drama classes   31 _____

32 What is the product of $-\frac{7}{8}$ and $-\frac{2}{14}$?

A $\frac{1}{4}$

B $\frac{1}{8}$

C $-\frac{1}{6}$

D $-\frac{9}{22}$ 32 _____

33 What is the probability that the Earth will make one complete revolution in the next 12 hours?

A 1

B 0.5

C 0.25

D 0   33 _____

34 A home building contractor bought $3\frac{1}{3}$ acres of land for $127,000. What was the cost of each acre?

A $38,100

B $45,000

C $42,000

D $37,100   34 _____

35 Look at the spinner. Which tally chart shows the most likely results of 22 spins of the spinner shown?

Letter	Spin Results
Q	IIII
R	IIII I
S	IIII
T	IIII
U	IIII

A

Letter	Spin Results
Q	II
R	IIII II
S	IIII
T	IIII
U	IIII

C

Letter	Spin Results
Q	IIII I
R	III
S	III
T	IIII
U	IIII I

B

Letter	Spin Results
Q	IIII
R	II
S	IIII
T	IIII III
U	III

D

35 _____

36 What is the area of the circle pictured in terms of π?

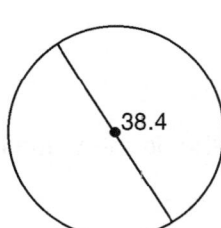

A 1,474.56π

B 286.64π

C 1,198.46π

D 368.64π 36 _____

Page 64   Test 5 – Part 1   Copyright © 2016
Topical Review Book Company

37 Becky is conducting an experiment where she times how long it takes 15 different dogs to run through a maze. The following histogram shows her results.

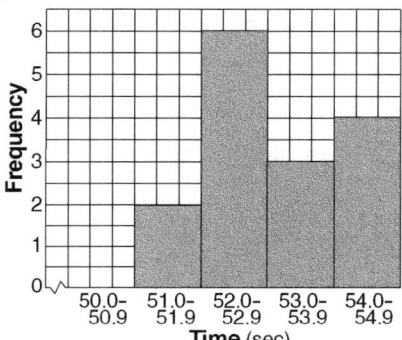

If Becky randomly selects one of the dogs to run the maze again, what is the probability that she choose a dog that runs the maze in less than 52.0 seconds?

A $\frac{8}{15}$

B $\frac{6}{15}$

C $\frac{2}{15}$

D $\frac{4}{15}$ 37 _____

38 Charlie rolls a six-sided die and flips a coin. Find the probability that Charlie's die and coin will land in the manner described:

| a number less than 3 and a head |

A $\frac{1}{2}$

B $\frac{1}{6}$

C $\frac{1}{3}$

D $\frac{1}{4}$ 38 _____

39 If 2 cards are dealt randomly from a standard deck of 52 cards, what is the probability that they will not be two face cards (king, queen, jack)?

A $\frac{40}{52} + \frac{39}{52}$

B $\frac{40}{52} + \frac{39}{51}$

C $\frac{40}{52} \cdot \frac{39}{51}$

D $\frac{40}{52} \cdot \frac{39}{52}$ 39 _____

40 Simplify the given expression: $9x + 4x =$

A $36x$

B $13x$

C $13x^2$

D 13 40 _____

41 How is the given expression rewritten using the Distributive Property?

| $0.7z + 2.3z$ |

A $z(0.7 + 2.3)$

B $0.7z(2.3z)$

C $z(0.7 - 2.3)$

D $2.3z(0.7z)$ 41 _____

Test 5 – Part 1

42 Two more than the product of four and a number is sixteen. Find the number.

A   −6
B $4\frac{1}{2}$
C   6
D $3\frac{1}{2}$ 42 _____

43 What inequality is equivalent to $\frac{3x}{2} - 6 < 9$?

A $x < 10$
B $x < 8$
C $x < 2$
D $x < 7$ 43 _____

44 Add: −15 + (−38)

A   +43
B   +53
C   −53
D   −23   44 _____

45 The diagram shows a right circular cone with four possible cutting paths, A, B, C, and D. What two-dimensional shape will result from a cross-section along path B in the given diagram?

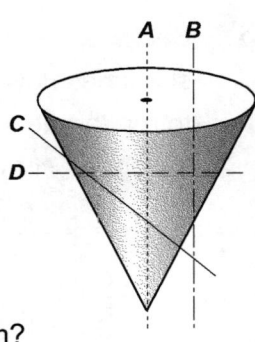

A   circle
B   parabola
C   triangle
D   oval   48 _____

46 What is the greatest common factor of 84 and 28?

A   12
B   28
C   4
D   14   46 _____

47 The measures of two angles of a triangle are 70° and 55°. This triangle is

A   a right triangle.
B   an obtuse triangle.
C   an equilateral triangle.
D   an isosceles triangle.   47 _____

Test 5 – Part 1

48 Camille is playing a game in the car with her brother, Cory. With the backs of the cards facing them both, Cory picks a card at random, notes the suit, and then returns the card to his sister's hand. Camille then has to guess the drawn card's suit.

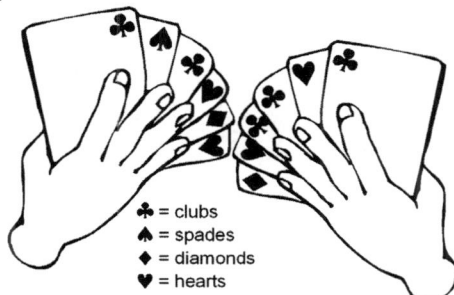

If Cory repeats the process described 144 times, predict how many times you would expect him to pick a heart or diamond card.

A   24
B   84
C   72
D   48

48 _____

49 The diagram represents a cube whose corner has been sliced off.

What is the sum of the interior angles of the two-dimensional surface resulting from this cross-section?

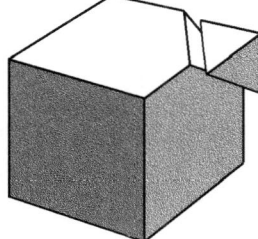

A   360°
B   720°
C   180°
D   90°

49 _____

50 What are a pair of adjacent angles in the diagram?

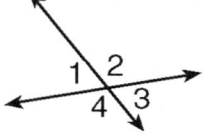

A   2 and 3
B   2 and 4
C   1 and 3
D   3 and 1

50 _____

51 What is the area of the figure?

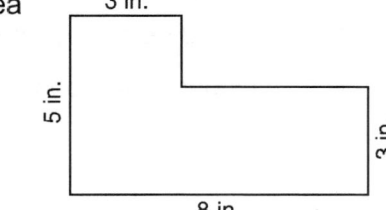

A   40 sq in.
B   26 sq in.
C   9 sq in.
D   30 sq in.

51 _____

52 Which one of the following tables shows a proportionality constant of 2?

A
x	y
2	4
3	5
4	6

C
x	y
2	4
3	3
4	2

B
x	y
2	4
3	9
4	16

D
x	y
2	4
3	6
4	8

52 _____

53 Four hundred licensed drivers participated in the math club's survey on driving habits. The table below shows the number of drivers surveyed in each age group.

Ages of People in Survey on Driving Habits

Age Group	Number of Drivers
16–25	150
26–35	129
36–45	33
46–55	57
56–65	31

Which statement best describes a conclusion based on the data in the table?

A It may be biased because no one younger than 16 was surveyed.

B It may be biased because the majority of drivers surveyed were in the younger age intervals.

C It would be fair because the survey was conducted by the math club students.

D It would be fair because many different age groups were surveyed.

53 _____

___

54 A car travels $16\frac{2}{3}$ miles on each gallon of gasoline. How many miles can the car travel on $14\frac{7}{10}$ gallons of gasoline?

A 224

B 245

C 264

D 235

54 _____

55 Find the product of the given set of numbers:

$$\left(+\frac{1}{7}\right)(+35)$$

A −5

B −7

C 5

D 7

55 _____

PART 2

56 30% of 45 is what number?

Show your work:

Answer: _____ months

57 Zechariah is playing games at the town carnival. Using data shown in the graph above, write an equation relating the number of games played to the number of tokens won.

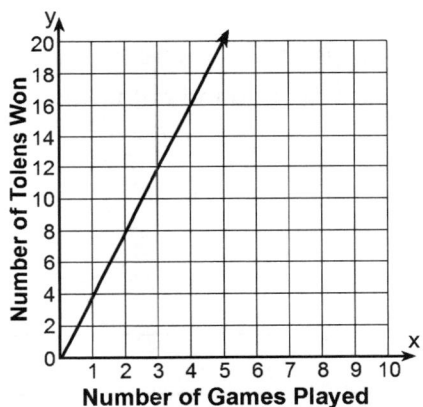

Answer: _____ months

58 7 students at Ryan's lunch table like cheese pizza and 9 students like pepperoni pizza. 4 students like both cheese and pepperoni pizzas. How many students like cheese pizza or pepperoni pizza or both? Use the Venn diagram below to support your answer.

Answer: _____ students

59 Tonya works a math problem on the board. She finds the surface area of the figure to be 24 in². Is her answer correct?

Show your work and explain your answer on the lines below:

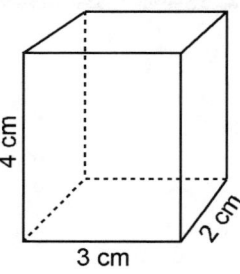

_____
_____
_____
_____

60 Bill and Nancy are riding a roller coaster.

Use the graph to determine the meaning of point *M* in relation to the data represented.

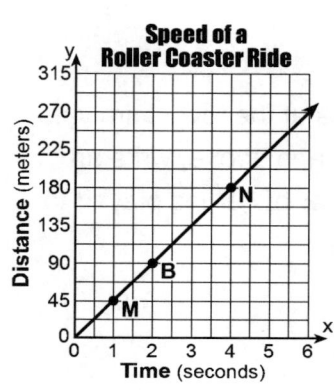

_____
_____
_____
_____
_____

61 Marcy baked 352 cookies for a community bake sale. If $\frac{7}{8}$ of the cookies are purchased, how many remaining cookies can Marcy donate to a retirement home?

Show your work:

Answer: _____ cookies

62 Josette enjoys making jewelry and is at a local craft store buying more supplies. She purchases a bag of silver beads for $12.99, a bag of purple beads for $7.49, and a spool of jewelry chain for $15.50. If she has a coupon for 40% off the highest price item, what is the total she will owe the craft store after 8% tax has been calculate into the price?

Show your work:

Answer: $_____

63 According to the Venn diagram below, what is the probability (to the nearest hundredth) of randomly choosing a student that plays both basketball and baseball?

Show your work:

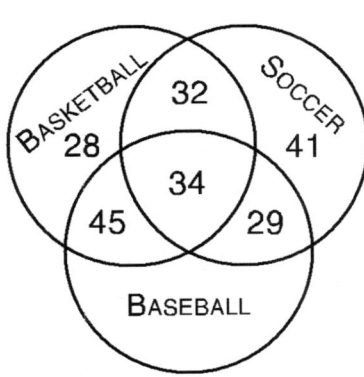

Answer: _____

64 In the accompanying diagram, parallel lines $\overline{AB}$ and $\overline{CD}$ are intersected by transversal $\overline{EF}$ at G and H, respectively.

If m∠CHG = (x + 20)° and m∠DHG = (3x)°, find the value of x.
Show your work:

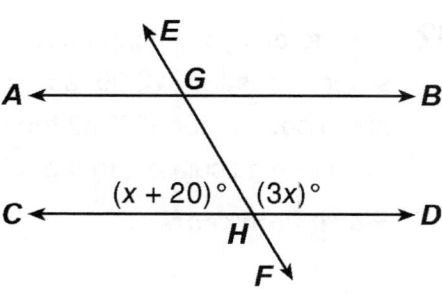

Answer: _____

65 Heather is at an Internet cafe. It costs her an initial charge of $1.25 plus an additional $0.10 each minute she uses their wireless connection to browse the Internet.

Part A
Write and solve an inequality that shows the maximum number of minutes Heather can use the wireless Internet connection with $5 to spend. Let x = the number of minutes she can use the wireless connection.
Show your work:

Inequality: _____

Answer: _____ minutes

Part B
Graph the solution set for the inequality you wrote in Part A on the number line below. Be sure you show that you can not browse fewer than zero minutes.

Test 6

Part 1

1 From a standard deck of 52 cards, one card is drawn. What is the probability that it will be a red card?

- A $\frac{4}{52}$
- B $\frac{2}{13}$
- C $\frac{26}{52}$
- D $\frac{13}{52}$

1 _____

2 Which one of the following expressions has the least value?

- A $(2 - 8) \times (-4 - 6)$
- B $(2 - 8) \times (-4) - 6$
- C $2 - (8 \times -4 - 6)$
- D $2 - 8 \times (-4) - 6$

2 _____

3 Solve the equation for the given variable:

$$3(4x - 2) = 8(x - 6)$$

- A $13\frac{1}{2}$
- B $-10\frac{1}{2}$
- C $10\frac{1}{2}$
- D $-13\frac{1}{2}$

3 _____

4 A mechanic estimates that a tire still has 8,000 miles of useful tread life. This is 20% of the original useful tread life of the tire. What was the tire's original useful tread life?

- A 40,000 miles
- B 50,000 miles
- C 1,600 miles
- D 30,000 miles

4 _____

5 Which of the following is a valid proportion?

- A $\frac{3}{2} = \frac{16}{32}$
- B $\frac{12}{8} = \frac{40}{30}$
- C $\frac{5}{12} = \frac{20}{60}$
- D $\frac{7}{5} = \frac{35}{25}$

5 _____

6 Simplify the given expression:

$$7y^2 + 3y^2 + 2y^2 =$$

- A $12y^8$
- B $12y^2$
- C $12y^6$
- D $12y$

6 _____

7 At a little league game, $880 was collected for hotdogs, hamburgers, and soda. All three items sold for $1.00 each. Three times as many hotdogs were sold as hamburgers. Four times as many sodas were sold as hamburgers. How many total hotdogs were sold?

A 330
B 110
C 300
D 440   7 _____

8 A sailboat is purchased for $7,250. A down payment of 18% is required. How much is the down payment?

A $1,350
B $1,205
C $1,305
D $5,945   8 _____

9 An appliance store sold 165 of the 300 food processors it had in stock. What percent of the food processors in stock did the company sell?

A 55%
B 45%
C 18.1%
D 35%   9 _____

10 Simplify the given expression:
$$-b^2 + 4b^2 =$$

A $3b^4$
B $-3b^2$
C $4b^2$
D $3b^2$ 10 _____

11 Which of the following is a valid proportion?

A $\frac{5}{6} = \frac{30}{38}$
B $\frac{25}{90} = \frac{5}{18}$
C $\frac{3}{51} = \frac{9}{152}$
D $\frac{11}{14} = \frac{50}{62}$ 11 _____

12 32% of 80 is what?

A 2,560
B 250
C 25.6
D 2.56   12 _____

13 Which of the following is a valid proportion?

A $\frac{10}{30} = \frac{60}{100}$

B $\frac{8}{5} = \frac{96}{60}$

C $\frac{2}{9} = \frac{60}{280}$

D $\frac{3}{4} = \frac{14}{20}$

13 _____

14 Which of the following is a valid proportion?

A $\frac{3}{4} = \frac{70}{100}$

B $\frac{6}{7} = \frac{33}{44}$

C $\frac{2}{10} = \frac{40}{120}$

D $\frac{13}{15} = \frac{117}{135}$

14 _____

15 Find the mean of the following numbers:

20, 32, 27, 28, 35, 32

A 32

B 30

C 29

D 26

15 _____

16 Solve the given expression: $-\frac{2}{3} + -\frac{2}{5}$

A $-\frac{16}{15}$

B $-\frac{4}{15}$

C $\frac{16}{15}$

D $-\frac{1}{2}$

16 _____

17 Given $x = 20$ and $y = 5$.

- The measure of angle $R = 2x + 4y$.
- The measure of angle $S = 5x + 4y$.

These two angles must be _____ angles.

A corresponding

B congruent

C complementary

D supplementary

17 _____

18 7% of 22 is what?

A 1.54

B 15.4

C 154

D 0.154

18 _____

Test 6 – Part 1   Page 75

19 Find the product of the given set of numbers:

$$(-8)(0)(-15)$$

- A 0
- B −15
- C −120
- D 120

19 _____

20 Simplify the given expression:

$$4y + (-2y) =$$

- A $2y^2$
- B −2y
- C 2y
- D 6y

20 _____

21 Paulie's Pet Store is having a "$\frac{1}{4}$ off" sale on parrots. What is the sale price of a parrot regularly priced at $168?

- A $100.80
- B $126
- C $84
- D $42

21 _____

22 Terrell is a landscaper for a local contractor. The table below shows the number of bushes needed to landscape around the lawns of different numbers of houses.

Bushes Needed for Landscaping

Number of Houses	2	5	7
Number of Bushes	18	45	63

Given the equation $y = 9x$ to represent the relationship between the number of houses (x) and the number of bushes (y) needed, what is the constant of proportionality?

- A 2
- B x
- C 18
- D 9

22 _____

23 What is 9 divided by $-\frac{2}{7}$?

- A $-\frac{18}{7}$
- B $\frac{2}{63}$
- C $-\frac{63}{2}$
- D $\frac{18}{7}$

23 _____

24 What is the value of $-4 \cdot (-8 + -3)$ when reduced to simplest form?

- A −44
- B 44
- C −20
- D 20

24 _____

25 The prices of the six textbooks purchased by a college student were $22.44, $31.29, $14.60, $29.40, $25.27, and $23.39. Find the median price.

A $24.33
B $24.40
C $22.44
D $23.39

25 _____

26 The diagram below shows a right circular cylinder with four possible cutting paths, 1, 2, 3, and 4.

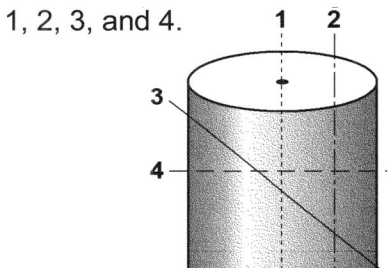

Which of the following statements best describes the difference in two-dimensional shapes resulting from cross-sections along path 1 and path 2 in the given diagram?

A   Cross-sections 1 and 2 are exactly the same shape and size.
B   Cross-section 2 is narrower than cross-section 1.
C   Cross-section 2 is a dilation of cross-section 1.
D   Cross-section 2 has greater surface area than cross-section 1.

26 _____

27 What is the value of x when m∠PQR = 80°?

A   10
B   7
C   9
D   6

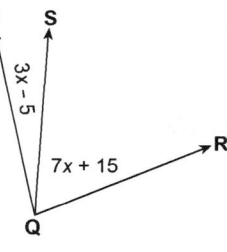

27 _____

28 What are a pair of adjacent angles in the diagram?

A s and t
B s and u
C x and z
D w and x

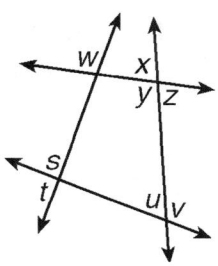

28 _____

29 What percent of 80 is 24?

A   0.3%
B   3%
C   40%
D   30%

29 _____

30 What is the volume of the following prism?

A   120 m³
B   200 m³
C   160 m³
D   180 m³

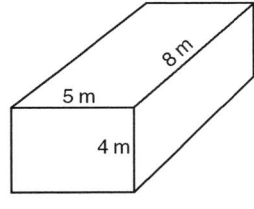

30 _____

Test 6 – Part 1

31 Mr. Gianni is in charge of determining what new game at the school carnival would be most popular.
Which would be the best sampling method?

A Survey all the neighbors that live around the school.
B Survey all the members of the football team.
C Survey two homerooms from each grade level at the school.
D Survey each homeroom in the 8th grade.   31 _____

32 What is $-\frac{3}{5}$ multiplied by $-\frac{6}{15}$?

A $-\frac{3}{25}$
B $\frac{18}{25}$
C $-\frac{18}{25}$
D $\frac{6}{25}$ 32 _____

33 A person standing at the chalkboard drops a piece of chalk. What is the probability that the chalk will not fall?

A 1%
B 100%
C 0%
D 50%   33 _____

34 A car travels $16\frac{2}{3}$ miles on each gallon of gasoline. How many miles can the car travel on $14\frac{7}{10}$ gallons of gasoline?

A 245
B 235
C 264
D 224   34 _____

35 Given the formula for the circumference of a circle.

Circumference = $2r\pi$

What is the circumference of the circle pictured?

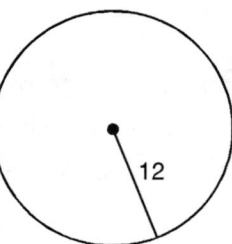

A 74.26
B 75.36
C 37.68
D 37.48   35 _____

36 In the diagram, angle 9 must equal 90°.

Which one of the following reasons proves this statement?

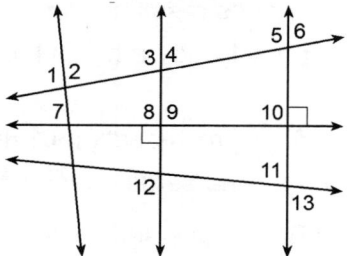

A Angle 9 and angle 10 are vertical angles.
B Angle 9 and angle 4 are vertical angles.
C Angle 9 and right angle are vertical angles.
D Angle 9 and angle 8 are vertical angles   36 _____

Page 78 Test 6 – Part 1

37 In the accompanying diagram, line $\overleftrightarrow{AB}$ and line $\overleftrightarrow{CD}$ intersect at point E.

If m∠AEC = (5x − 22)° and m∠DEB = (2x + 35)°, what is m∠AED?

A 107°
B 73°
C 19°
D 95°

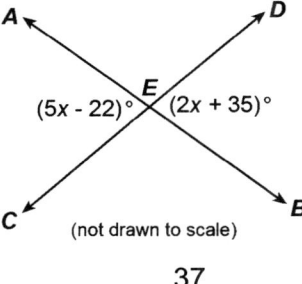
(not drawn to scale)

37 _____

38 What is the volume of the cube shown?

A $1\frac{1}{2}$ mm³

B $\frac{1}{8}$ mm³

C $\frac{1}{16}$ mm³

D $\frac{1}{6}$ mm³

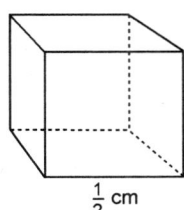
$\frac{1}{2}$ cm

38 _____

39 Simplify the given expression: 7c + 14c =

A $2c^2$
B 7c
C 21
D 21c

39 _____

40 Ms. Malone divides her class into 4 teams. Their challenge is to find the best test to predict the probability of rolling a prime number with a standard die, numbered 1 through 6. Which one of the following teams has the best plan to get the most accurate results?

A Team D — roll the die 10 times and record how many times the numbers 3 or 5 land face up

B Team C — roll the die 100 times and record how many times the numbers 3 or 5 land face up

C Team A — roll the die 10 times and record how many times the numbers 2, 3, or 5 land face up

D Team B — roll the die 100 times and record how many times the numbers 2, 3, or 5 land face up

40 _____

41 Kathryn is buying her wedding dress. The dress is on sale for 35% of the regular price (w). What expression represents the sale price of the wedding dress?

A 1.35w
B 1.65w
C 0.65w
D 0.35w

41 _____

42 What is the surface area of the solid?

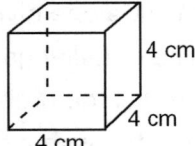

A 64 cm²
B 24 cm²
C 46 cm²
D 96 cm²

42 _____

43 What inequality is represented by the graph below?

A $x \geq -3$
B $x < -3$
C $x > -3$
D $x \leq -3$

43 _____

44 What is the length of the line segment joining points X and Y on the graph below?

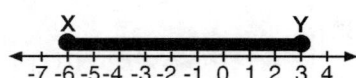

A 3
B 10
C 9
D 6

44 _____

45 Which of the following three-dimensional sketches matches the different views of the solid below?

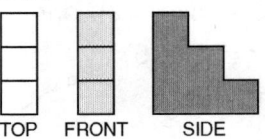

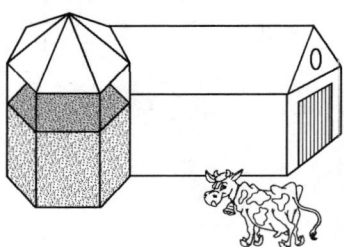

45 _____

46 The drawing below shows a view of a barn with an attached silo.

Which of the following drawings best represents the top view of this building?

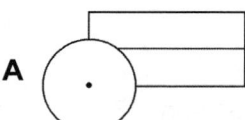 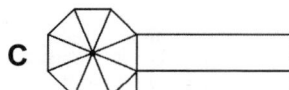

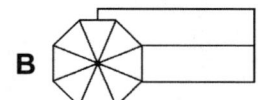 

46 _____

Page 80 — Test 6 – Part 1

47 The diagram below represents a cube that has been cut along a diagonal plane.

What is the shape of the two-dimensional surface that results from this cross-section?

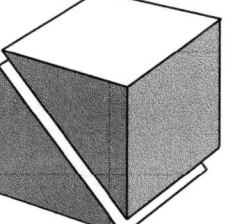

- A right triangle
- B square
- C rectangle
- D diamond

47 _____

48 Given the formula for the circumference of a circle.

$$\text{Circumference} = 2\pi r$$

What is the circumference of the circle pictured?

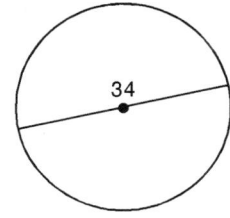

- A 34π
- B 68π
- C 17π
- D 43π

48 _____

49 The drawing represents a 3-dimensional solid.

What geometric shape best describes this solid when viewed from the top?

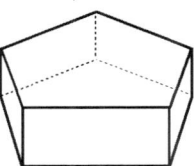

- A octagon
- B pentagon
- C decagon
- D hexagon

49 _____

50 What are a pair of adjacent angles in the diagram?

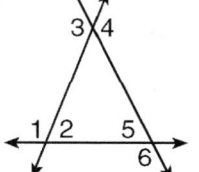

- A 5 and 6
- B 1 and 3
- C 3 and 4
- D 2 and 5

50 _____

51 What is the area of the triangle below?

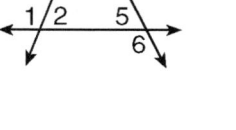

- A 12 units²
- B 8 units²
- C 13.5 units²
- D 9 units²

51 _____

52 Which one of the following tables shows a proportionality constant of 3?

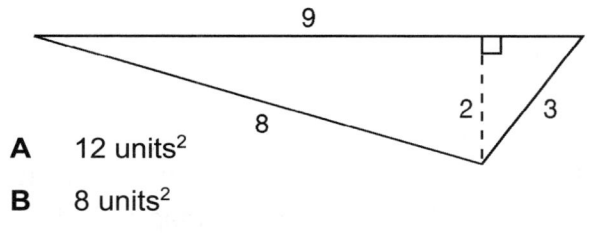

	x	y
A	3	9
	4	12
	5	15
	6	18

	x	y
C	3	12
	4	16
	5	20
	6	24

	x	y
B	3	6
	4	8
	5	10
	6	12

	x	y
D	3	6
	4	12
	5	20
	6	30

52 _____

Test 6 – Part 1

53 A survey is being conducted to determine which types of television programs people watch. Which survey and location combination would likely contain the most bias?

    A randomly surveying 75 people during the day in a clothing store

    B surveying the first 25 people who enter a grocery store

    C surveying 10 people who work in a sporting goods store

    D randomly surveying 50 people during the day in a mall   53 _____

54 If the actual distance between two locations is 48.8 kilometers and the map scale is 3 centimeters = 24 kilometers, what is the distance between the two locations on the map?

    A 0.7 cm

    B 8.1 cm

    C 390.4 cm

    D 6.1 cm   54 _____

55 Ashley earns $140 per week plus 10% commission on all sales. How much were her sales if she made $375 in one week?

    A $515.00

    B $2,350.00

    C $177.50

    D $3,890.00   55 _____

PART 2

56 Pam opened a bank account with a $200 deposit. The account had a simple interest rate of 4.6%. Her most recent bank statement showed a balance of $328.80. If she made no deposits or withdrawals, how long ago was the account opened?

Show your work:

Answer: _____ years

57 Chad is training for a race. He is going to run a total of 8 miles around the track. Each lap of the track is a $\frac{1}{2}$ mile. How many total laps did he run?

Show your work:

Answer: _____ years

58 **Part A**

Belinda wants to purchase six apps priced at $0.99 each and the remaining at the normal pricing of $2.75 each. Write and use estimation to solve an inequality showing the maximum number of apps (*a*) Belinda can purchase with a $50 gift card. Show your work.

Show your work:

Inequality: _____

Answer: _____

Part B

Explain why your estimated result is more or less than the actual result.

_____
_____
_____
_____
_____

59 Find the volume of a cube having the edge measures shown below.

Show your work:

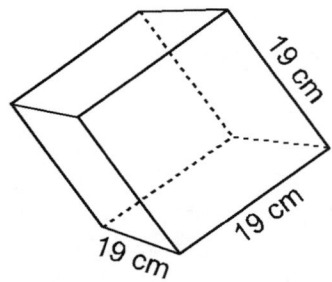

Answer: _____ mm³

60 On average, a certain baseball player hits a baseball 1 time in every 4 swings of the bat. Suppose he makes exactly 520 swings during one season.

Part A

Explain the simulation you would create to estimate the experimental probability of the number of hits the player made that season.

_____
_____
_____
_____
_____

Part B

How would you alter your simulation in Part *A* to represent that 2 out every 13 hits of the baseball results in a foul ball?

_____
_____
_____
_____
_____

61 The graph shows the distances of three remote control cars over time when racing on a track. Are the graphed lines of each car's speed proportionate to each other?

Explain your reasoning:

_____
_____
_____
_____
_____
_____
_____

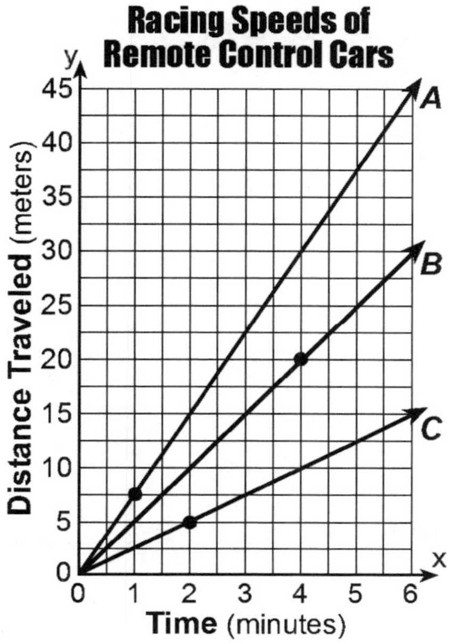

Racing Speeds of Remote Control Cars

62 In the space below, use models to show how $6 \div \frac{1}{2}$ is equivalent to 6×2.

63 A receipt shows the purchase of two items: a coffee maker for $49.99 and a coffee mug for $15.99. The customer had two coupons, shown below.

Which coupon would she have used to save the most amount of money?

Show your work:

Coupon: _____

64 In the diagram below, four lines intersect to create fourteen angles.

In the given diagram, angle 11 measures $(5x - 35)°$ and angle 12 measures $(9x + 5)°$. Solve for x.

Show your work:

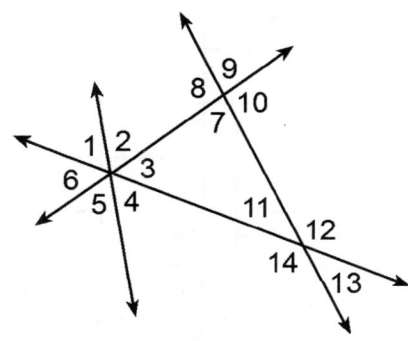

Answer: _____

65 A middle school is adding a vending machine to the cafeteria. The administration decides to conduct a poll to determine the types of food and/or drinks they will supply. Decide if the given poll sample is representative or biased.

Justify your reasoning.

Poll two school buses of students.

_____
_____
_____
_____

Correlation of Standards

QUESTION	TEST 1	TEST 2	TEST 3	TEST 4	TEST 5	TEST 6
1	7.RP.2a	7.RP.1	7.RP.2a	7.RP.3	7.SP.C.8a	7.SP.C.8a
2	7.RP.2b	7.RP.2a	7.EE.B.4b	7.EE.B.4b	7.NS.1d	7.NS.3
3	7.RP.2b	7.RP.3	7.RP.2b	7.RP.1	7.EE.B.4a	7.EE.A.1
4	7.RP.2b	7.SP.8b	7.RP.2a	7.RP.2d	7.RP.3	7.RP.3
5	7.RP.2b	7.NS.1a	7.EE.A.1	7.EE.A.1	7.RP.2b	7.RP.2a
6	7.RP.2b	7.SP.7b	7.RP.2d	7.RP.2b	7.EE.A.1	7.EE.A.1
7	7.RP.2a	7.RP.3	7.RP.2a	7.RP.2b	7.EE.B.4a	7.EE.B.3
8	7.RP.3	7.NS.2a	7.RP.3	7.RP.3	7.RP.3	7.RP.B.3
9	7.RP.2b	7.EE.A.1	7.NS.1b	7.NS.1d	7.RP.3	7.RP.3
10	7.RP.3	7.RP.2d	7.NS.1d	7.NS.1d	7.EE.A.1	7.EE.A.1
11	7.RP.3	7.RP.1	7.NS.3	7.NS.3	7.RP.2a	7.RP.2a
12	7.RP.3	7.NS.2b	7.EE.A.2	7.EE.A.2	7.RP.3	7.RP.3
13	7.RP.3	7.EE.A.1	7.G.A.1	7.G.A.1	7.RP.1	7.RP.2a
14	7.RP.3	7.G.B.5	7.G.A.2	7.G.A.2	7.RP.3	7.RP.2a
15	7.NS.1c	7.G.A.3	7.G.B.5	7.RP.2a	7.RP.2d	7.SP.B.3
16	7.NS.1c	7.G.B.5	7.SP.A.2	7.G.A.1	7.NS.1d	7.NS.1d
17	7.NS.1c	7.RP.1	7.RP.1	7.RP.1	7.SP.C.8a	7.G.B.5
18	7.NS.1d	7.RP.3	7.SP.B.4	7.SP.A.2	7.RP.3	7.RP.3
19	7.NS.1c	7.SP.A.1	7.G.B.6	7.G.B.6	7.NS.2a	7.NS.2a
20	7.NS.2d	7.G.B.6	7.SP.A.1	7.SP.A.1	7.EE.A.1	7.EE.A.1
21	7.NS.2d	7.NS.2a	7.SP.A.2	7.SP.A.2	7.RP.2d	7.RP.3
22	7.NS.2d	7.SP.B.4	7.SP.C.8b	7.SP.C.8b	7.RP.1	7.RP.2d
23	7.RP.2a	7.RP.3	7.SP.C.7a	7.SP.B.4	7.NS.2c	7.NS.2c
24	7.RP.2a	7.NS.1b	7.SP.B.4	7.SP.C.6	7.NS.3	7.NS.3
25	7.RP.2a	7.SP.C.5	7.NS.3	7.RP.2a	7.G.A.2	7.SP.B.3
26	7.RP.2a	7.NS.3	7.EE.A.1	7.EE.A.1	7.G.A.3	7.G.A.3
27	7.RP.2a	7.SP.C.6	7.EE.B.4b	7.EE.B.4b	7.G.B.5	7.G.B.5
28	7.EE.B.3	7.SP.C.7b	7.EE.B.4a	7.EE.B.4a	7.RP.1	7.G.B.5
29	7.EE.A.1	7.G.B.4	7.EE.B.4a	7.SP.C.5	7.RP.3	7.RP.3
30	7.EE.A.1	7.SP.C.7b	7.SP.C.5	7.SP.C.5	7.G.B.6	7.G.B.6
31	7.EE.A.1	7.SP.C.8c	7.SP.C.5	7.SP.C.5	7.SP.A.1	7.SP.A.1
32	7.EE.A.1	7.SP.C.7a	7.RP.2a	7.RP.2a	7.NS.2c	7.NS.2c
33	7.EE.A.1	7.RP.2c	7.RP.3	7.RP.3	7.SP.C.5	7.SP.C.5